# Technological Change

**Studies in The History of Science, Technology and Medicine**
*Edited by John Krige, European University Institute, Florence, Italy*

**Series Introduction**

*Studies in the History of Science, Technology and Medicine* is a new book series which aims to stimulate research in the field, concentrating on the twentieth century. It seeks to contribute to our understanding of science, technology and medicine as it is embedded in society, exploring the links between these subjects on the one hand and the cultural, economic, political and institutional contexts of their genesis and development on the other. Within this framework, and while not favouring any particular school of methodological approach, it welcomes studies which examine relations between science, technology, and society in new ways, e.g. the social construction of technologies, large technical systems.

**Volume 1**
Technological Change: Methods and Themes in the History of Technology
*Edited by Robert Fox*

**Volume 2**
Technology Transfer out of Germany after 1945
*Edited by Matthias Judt and Burghard Ciesla*

**Volume 3**
Entomology, Ecology and Agriculture: The Making of Scientific Careers in North America, 1885–1985
*Paolo Palladino*

**Volume 4**
The Historiography of Contemporary Science and Technology:
Whose History? Whose Science?
*Edited by Thomas Söderqvist*

**Volume 5**
Science and Spectacle: The Work of Jodrell Bank in Postwar British Culture
*Jon Agar*

**Volume 6**
Molecularising Biology and Medicine: New Practices and Alliances, 1910s–1970s
*Soraya de Chadarevian and Harmke Kamminga*

**Other Volumes in Preparation**

Making Isotopes Matter: F.W. Aston and the Culture of Physics
*Jeff Hughes*

This book is part of a series. The publisher will accept continuation orders which may be cancelled at any time and which provide for automatic billing and shipping of each title in the series upon publication. Please write for details.

# Technological Change

## Methods and Themes in the History of Technology

*Edited by*
*Robert Fox*
*University of Oxford, UK*

Routledge
Taylor & Francis Group

LONDON AND NEW YORK

First Published 1996
Second Printing 1998

Published by Routledge
2 Park Square, Milton Park, Abingdon, Oxon, OX14 4RN
270 Madison Ave, New York NY 10016

Transferred to Digital Printing 2009

---

**British Library Cataloguing in Publication Data**

Technological Change: Methods and Themes
in the History of Technology. — (Studies
in the History of Science, Technology &
Medicine, ISSN 1024-8084; Vol. 1)
  I. Fox, Robert   II. Series
  609

  ISBN 3-7186-5792-9

FRONT COVER          Johann Faulhaber, *Ingenieurs-Schul. Erster Theil*
                              (Frankfurt-am-Main, 1630).
                    By permission of The British Library 8548.b. 11 (14)

**Publisher's Note**
The publisher has gone to great lengths to ensure the quality of this reprint but points out that some imperfections in the original may be apparent.

# Contents

# Preface

The contributions to this volume are texts that have been reworked and extended after their initial presentation as papers at the conference on "Technological change", held in Oxford between 8 and 11 September 1993. I wish to express my thanks to the main sponsors of the conference — the Modern History Faculty of the University of Oxford and the Royal Society — and to the Renaissance Trust, which gave generous financial support. I am also grateful for a grant from the International Union of the History and Philosophy of Science and to the Warden and Trustees of Rhodes House for making their gracious premises available for the conference.

The degree of international collaboration in the conference was a source of particular satisfaction, and it is a pleasure to acknowledge the varied contributions that were made, at the stages of both planning and realization, by partners in France (Cité des Sciences et de l'Industrie/CNRS, Paris), Germany (Deutsches Museum, Munich), and Italy (University of Bologna), as well as by the British Society for the History of Science and the National Museum of Science and Industry in Britain. The BSHS made an especially important contribution by placing the services of its Executive Secretary, Wing Commander Geoffrey Bennett, at the disposal of the conference. The smooth running of the conference also owed much to Priscilla Frost, to an energetic local committee, and to the resourcefulness, energy, and good humour of an unfailingly helpful team of graduate students.

In preparing the texts for publication, I have been greatly helped by the meticulous and efficient work of Stephanie Jenkins in the Modern History Faculty. I extend my warmest thanks to her, as I do to all the contributors to the conference and to this book, which I am happy to see included in Harwood Academic's growing list in the history of technology.

ROBERT FOX

# Introduction

## Methods and Themes in the History of Technology

### Robert Fox

The essays in this volume arise from papers given at the conference on "Techno-logical change" at Rhodes House and Wadham College, Oxford, from 8 to 11 September 1993. Quite deliberately, the conference was conceived as a sequel to an earlier one, on "The structure of scientific change", that was organized by Alistair Crombie in Oxford in 1961.[1] The connexion between the two gatherings was, in certain obvious respects, a distant one. But in each case there was an attempt both to review current trends in historiography and to give an impression of empirical work in some of the most active areas of contemporary research. Selecting the fourteen contributions in this volume from over a hundred papers that were delivered at the 1993 conference was a difficult task. But those that appear here reflect rather faithfully, if not comprehensively, the interaction between theoretical and empirical concerns which the conference sought to promote.

Another point of similarity between the conferences of 1961 and 1993 was that they occurred at periods of fundamental debate about methodology. In the early 1960s, the history of science was beginning to make the more intimate links with other disciplines — in particular in the social sciences — that have set the tone of historiography ever since. One mark of this, at the Oxford con-ference, was Thomas Kuhn's paper on "The function of dogma in scientific

---

[1] For the papers that were delivered at the conference, see A.C. Crombie (ed.), *Scientific change. Historical Studies in the Intellectual, Social and Technical Conditions for Scientific Discovery and Technical Invention, from Antiquity to the Present. Symposium on the history of science. University of Oxford 9–15 July 1961* (London: Heinemann Educational Books, 1963).

research", a profoundly sociological study of the processes of scientific change that advanced ideas published a year or so later in *The structure of scientific revolutions*.[2] Thirty years on, the history of technology was engaged in a similar process of bridge-building, again mainly with the social sciences.

There were, however, important differences. Most obviously, the process was more advanced in the history of technology in the early 1990s than it had been in the history of science at the time of the first Oxford conference. One reason for this was that the history of technology lends itself particularly well to certain disciplinary alliances, most notably with economic and business history. Significantly, it was as an economic historian (though one who was many things besides) that Marc Bloch urged that greater attention should be paid to the history of technology in the 1930s and began his own forays into the field. Then Bloch's call, launched with Lucien Febvre in the young *Annales d'histoire économique et sociale*, went largely unheeded.[3] But eventually, and especially since the 1950s, the work of economists and economic historians has come to be seen as highly relevant to the history of technology. Along the way, there has been some resistance, chiefly from historians of technology concerned to preserve the integrity of their discipline and fearful that the content of the technologies they studied might be marginalized. But now, such theoretical notions as those of Joseph Schumpeter and Jacob Schmookler on the process of technological innovation, Nathan Rosenberg on the epistemological dangers of "black boxing" technology, and Paul David on path dependence and positive feedback have passed from the realms of economics and economic history to become part and parcel of the historian of technology's methodological armoury.[4] The results of this methodological enrichment are to be seen in case studies whose simultaneous treatment of both technical and economic issues shows how fruitful the process of assimilation has been. Following the pace-making work of

---

[2] T.S. Kuhn, "The function of dogma in scientific research", in Crombie, *Scientific Change*, pp. 347–69. The volume appeared as *The Structure of Scientific Revolutions* (Chicago: University of Chicago Press, 1962)

[3] Through reviews, editorials, and case studies in the *Annales* between the foundation of the journal in 1929 and the mid-1930s, Bloch and Febvre sought repeatedly to secure greater attention for the history of technology. See, in particular, the special issue devoted to the history of technology in November 1935, for which Febvre wrote a notable preface: "Réflexions sur l'histoire des techniques", *Annales d'histoire économique et sociale* (1935), **7**: 531–5.

[4] The texts in which the ideas of these authors have been expressed include: J.A. Schumpeter, *The Theory of Economic Development. An Inquiry into Profits, Capital, Credit, Interest, and the Business Cycle* (Cambridge, Mass.: Harvard University Press, 1934); J. Schmookler, *Invention and Economic Growth* (Cambridge, Mass.: Harvard University Press, 1956); N. Rosenberg, *Inside the Black Box. Technology and Economics* (Cambridge: Cambridge University Press, 1982); P.A. David, "Heroes, herds and hysteresis in technological history: Thomas Edison and 'The battle of the systems' reconsidered", *Industrial and Corporate Change* (1992), **1**: 129–80.

David Landes in the 1960s, Hugh Aitken, Merritt Roe Smith, Philip Scranton, Leonard Reich, David Hounshell, and John Kenly Smith have all demonstrated the particular vigour of this tradition in the United States.[5] while François Caron, Svante Lindqvist, and Ulrich Wengenroth are just three of a growing number of European scholars whose writing over more than a decade has spanned a similar disciplinary divide.[6]

Another reason for the different positions of the history of science in the early 1960s and the history of technology in the early 1990s was that historians of technology had long been able to draw on what had already been achieved in the history of science. Very pertinent lessons were learned, for example, from the attempts of historians of science to exploit ideas originating in the sociology of knowledge. The most influential work, stemming from the contributions of David Bloor and Barry Barnes in the 1970s, was that which portrayed scientific knowledge as, at least in some degree, "socially constructed". If, as Bloor, Barnes, and others working in the same tradition argued, the very content of science is fashioned by its social environment, might the content of technology (and not just its institutions and social role) be fashioned in the same way? Among the early explorations of this possibility, the study of the social construction of the bicycle by Trevor Pinch and Wiebe Bijker has become a classic, as has the volume in which it appeared, which Pinch and Bijker edited in 1987 with Thomas Hughes.[7] All subsequent work by both the exponents and the critics of the social construction of technology (or SCOT) approach has taken this volume as its bench mark.

---

[5] Among the works by these authors, see D.S. Landes, *The Unbound Prometheus. Technological Change and Industrial Development in Western Europe from 1750 to the Present* (Cambridge: Cambridge University Press, 1969); H.G.J. Aitken, *Syntony and Spark. The Origins of Radio* (New York: John Wiley & Sons, 1976); M. Roe Smith, *Harpers Ferry Armory and the New Technology. The Challenge of Change* (Ithaca, NY: Cornell University Press, 1977); P. Scranton, *Proprietary Capitalism. The Textile Manufacture at Philadelphia, 1800–1885* (Cambridge: Cambridge University Press, 1983); L.S. Reich, *The Making of American Industrial Research. Science and Business at GE and Bell, 1876–1926* (Cambridge: Cambridge University Press, 1985); D.A. Hounshell and J. Kenly Smith, *Science and Corporate Strategy. Du Pont R&D, 1902–1980* (Cambridge: Cambridge University Press, 1988).

[6] See, as examples of the work of these scholars, F. Caron, *Le résistible déclin des sociétés industrielles* (Paris: Perrin, 1985); S. Lindqvist, *Technology on Trial. The Introduction of Steam Power Technology into Sweden, 1715–1736* (Stockholm: Almqvist & Wiksell, 1984), and U. Wengenroth, *Enterprise and Technology. The German and British Steel Industries, 1865–1895*, translated [from the original German edition of 1986] by Sarah Hanbury Tenison (Cambridge: Cambridge University Press, 1994).

[7] T.J. Pinch and W.E. Bijker (eds.), "The social construction of facts and artifacts: or how the sociology of science and the sociology of technology might benefit each other", in W.E. Bijker, T.P. Hughes, and T.J. Pinch (eds.), *The Social Construction of Technological Systems. New Directions in the Sociology and History of Technology* (Cambridge, Mass.: MIT Press, 1987), pp. 17–50.

The opening of the discipline of the history of technology and the attendant proliferation of methodologies and historiographical alliances show no sign of abating. Indeed, a striking characteristic of a recent volume of essays edited by Wiebe Bijker and John Law is precisely the variety of the approaches that have come to be exploited, even within a group of scholars who share a commitment not just to the social "shaping" of technology but, more specifically, to a vision in which technology and society are treated as part and parcel of a single unified inquiry.[8] The same diversity was very much to the fore at the Oxford conference. Trevor Pinch's paper was one that brought out with particular clarity the many levels of possible commitment to what could still be described as a SCOT methodology.

At one end of Pinch's SCOT spectrum, as presented here in the published version of his paper, is a "mild" version, which would simply acknowledge the social forces (pressure groups, for example) that contribute to the final form of a technology. As Pinch admits, this version is vulnerable to the charge, levelled recently by David Edgerton, that SCOT offers little that was not already fully recognized by historians sensitive to social context.[9] At the other end of the spectrum, however, is a "radical" version that would extend a SCOT analysis to every detail of the content of a technology, even including the criteria by which the technology might be adjudged to be working. In this sense, Donald MacKenzie's study of the conflicting conclusions about the accuracy of missiles, both in his book *Inventing Accuracy*[10] and in his contribution to this volume, are firmly in the radical tradition. In answer to his question "How do we know the properties of artefacts?", MacKenzie's exploration of the disparity between different assessments of the effectiveness of the American Patriot missile during the Gulf War points to the central role of politically inspired perceptions in what, on the face of it, had the character of an unproblematic technological assessment. The debate that MacKenzie analyses was sparked off by Theodore Postol's challenge to the official judgement that Patriot had been a virtually unqualified success in its task of intercepting Scud missiles. Postol's argument, though strong, was not water-tight. But, viewed through the lens of the sociology of knowledge, Patriot's carefully filmed performance, on which its defenders based their case, emerges as highly contestable evidence; certainly, any notion that the weapon's performance was assessed by a dispassionate process of induction is, as a result, impossible to sustain.

Another strong tradition in the historiography of the last two decades has been that associated with the notion of a technological system, pioneered

[8] W.E. Bijker and J. Law (eds.), *Shaping Technology/Building Society. Studies in Sociotechnical Change* (Cambridge, Mass.: MIT Press, 1992).

[9] D. Edgerton, "Tilting at paper tigers", *The British Journal for the History of Science* (1993), **26**: 67–75.

[10] D. MacKenzie, *Inventing Accuracy. A Historical Sociology of Nuclear Missile Guidance* (Cambridge, Mass.: MIT Press, 1990).

independently by Bertrand Gille and Thomas Hughes. In fact, the conception of a system in the writing of Gille is very different from that deployed by Hughes, as Antoine Picon points out in this volume. And despite the mark that Gille has made in France, through his magisterial *Histoire des techniques* and the birth of the journal *Culture technique*,[11] it is Hughes's approach, applied in his *Networks of Power. Electrification in Western Society 1880–1930* in 1983, that is better-known in the English-speaking world.[12] In *Networks of Power*, as in all his writing, Hughes embeds technology as firmly in society as even the most radical work in the SCOT canon. He emphasizes the interconnectedness of the diverse elements constituting a large technological system, such as the electrical power supply networks that are treated in his book: on this model, an element that lags behind or gets out of step with the others in the system (a "reverse salient") will prevent the whole system from advancing and demand special attention before further development can continue.[13]

The other striking feature of Hughes's model, in both his early and his subsequent work, has been the emphasis on the extent and complexity of his systems. The system that allowed Thomas Edison's new incandescent electric lamp to triumph over its rivals in the 1880s embraced not only the lamp itself but also the associated technical elements, such as generators and cables, and (no less importantly) the financial, political, and demographic circumstances in the various contexts in which the invention was exploited.[14] Hughes's contribution to *Technological Change,* which provides a foretaste of his forthcoming book on the creation of large technological systems since 1945, suggests that the complexity of such systems has grown inordinately over the last hundred years and continues to do so. Indeed, it may well be that complexity has become, in Hughes's words, a "defining characteristic" of our whole post-modern world; in technology, as he argues, the complexity calls urgently for ever more effective forms of management, often "organization-transcending" forms such as the

---

[11] Gille's *Histoire des techniques* (Paris: Gallimard, 1978) appeared under his editorship in the Pléiade series. In fact, Gille was far more than an editor, writing a large part of the volume himself. An English translation appeared as *The History of Techniques*, translated by P. Southgate *et al.* (2 vols., New York: Gordon and Breach Science Publishers, 1986). The early issues of *Culture technique,* which began publication in 1979, bore the mark of Gille's influence, although his premature death in 1981 deprived the publication of his sustained engagement.

[12] In addition to Hughes's *Networks of Power* (Baltimore, Md: Johns Hopkins University Press, 1983), see some of his more recent work, notably his *American Genesis. A Century of Invention and Technological Enthusiasm 1870–1970* (New York: Viking, 1989).

[13] For the concept of the "reverse salient" and the associated idea of the "critical problem", see Hughes, *Networks of Power*, pp. 79–105, and his chapter "The dynamics of technological change: salients, critical problems, and industrial revolutions", in Giovanni Dosi, Renato Giannetti, and Pier Angelo Toninelli (eds.), *Technology and Enterprise in a Historical Perspective* (Oxford: Clarendon Press, 1992), pp. 97–118.

[14] Ibid., pp. 18–46.

interdisciplinary committees that were used during and immmediately after the second world war to coordinate the academic, industrial, and military facets of the United States' atomic weapons and air defence systems.

Although writings in the SCOT and the systems traditions have tended to command the lion's share of attention in recent historiographical debate, this has not prevented other approaches from blossoming as well. Three explicitly theoretical papers in this volume — by Joel Mokyr, John Pickstone, and Antoine Picon — reflect something of the richness of the methodological menu now available to historians. Mokyr's paper, which seeks to demonstrate the existence of a Darwinian logic at work in the process of technological change, is written primarily for the attention of his fellow-economic historians, a number of whom, following the work of R.R. Nelson and S.G. Winter in the late 1970s and early 1980s,[15] have explored the analogies between Darwin's notions of variation and selective retention and the mechanisms for the selection of successful technologies. In fact, evolutionary models of technological change have a much longer history than this, as S. Colum Gilfillan's remarkable pioneering studies, dating from sixty years ago, show.[16] But the renewed vogue for such models has not only given them new prominence but also extended their currency far beyond the realms of economic history in which, in recent years, they have tended to be particularly favoured.[17] Mokyr's cautionary reflexions on the snares inherent in the application of Darwinian principles should therefore make their mark in disciplinary communities other than the one he explicitly addresses.

Models, of course, can beguile and intrigue rather than instruct. To stay with the one that Mokyr treats, it is evident that the undisciplined invocation of evolutionary parallels between the worlds of biology and technology will yield nothing but a curious, elegant metaphor. Used with due attention to the

---

[15] In this, R.R. Nelson and S.G. Winter, *An Evolutionary Theory of Economic Change* (Cambridge, Mass.: Belknap Press of Harvard University Press, 1982), has been especially influential.

[16] S. Colum Gilfillan, *Inventing the Ship. A Study of the Inventions made in her History between Floating Log and Rotorship. A Self-contained but Companion Volume to the Author's "Sociology of Invention"* (Chicago: Follett Publishing Co., 1935), and *The Sociology of Invention. An Essay in the Social Causes, Ways and Effects of Technic Invention, especially as demonstrated historically in the Author's "Inventing the Ship"* (Chicago: Follett Publishing Co., 1935). The period from the late 1920s to the mid-1930s was rich in studies of the profile and character of technological change. In the years immediately preceding the publication of these books by Gilfillan, Abbott Payson Usher, William Fielding Ogburn, and Lewis Mumford had published, respectively, *A History of Mechanical Inventions*, *Living with Machines* (both incorporating evolutionary models), and *Technics and Civilization* (structured around Mumford's very different notion of successive eotechnic, paleotechnic, and neotechnic phases).

[17] For a recent use of an evolutionary model by an historian of technology, see George Basalla, *The Evolution of Technology* (Cambridge: Cambridge University Press, 1988).

intricacies of evolutionary theory, however, the metaphor is replete with exploitable insights. One strong message that Mokyr is able to deliver, for example, is that diversity and its inevitable corollary, wastefulness, are foundations, even essential foundations, for creativity: the analogy is with the random variations on which natural selection operates. Hence diversity and wastefulness on this model are anything but marks of economic inefficiency. They constitute the possibilities from which the inventive mind fashions its new, creative synthesis.[18]

Like Mokyr, both Pickstone and Picon explore ways of getting to grips with the "big picture" of the history of technology. Their emphasis is on the intellectual dimension of technological change, although both are committed to an intellectual history firmly rooted in the broader social and economic context and enriched by interdisciplinary sensibilities. Casting a bridge between the history of technology and the history of science and medicine, Pickstone constructs his argument from his engagement with Michel Foucault's conception of eighteenth-century natural history as part of a classical *episteme*. While rejecting the notion of *episteme* in his own analysis, Pickstone argues that during the Enlightenment, artefacts and practices in the craft tradition were viewed by those he terms "savant" commentators (in the *Encyclopédie*, for example) as products of particular, predominantly local circumstances; they were "types", each of them displaying a distinctive profile of characteristics that allowed them to be classified, in the same way as plants or animals. At about the time of the French Revolution (which also coincided with the collapse of Foucault's classical *episteme*), such "savant" ways of knowing came to be dominated by new analytical approaches, not only in engineering but also (in a way that is crucial to Pickstone's argument) in medicine and chemistry. In the specific contexts in which it appeared, the new approach brought out the universality of processes rather than their differentiated local character and stressed the common, constitutive elements, whether of machines, diseased bodies, or chemicals; all were now seen as "compounds". In seeking to explain the extension and the power of analysis in this period, Pickstone focusses on the new state institutions and pedagogical programmes of Paris and on the intensification of capitalist rationalization and industrialization in Britain. But he also wishes to underline the concurrence of his different "ways of knowing": analytical projects were already available as models (in planetary astronomy, for example), and savant or craft practices continued to be developed. Hence we are invited to

---

[18] This view of technological creativity suggests parallels with ideas advanced quite independently in Basalla, *The Evolution of Technology*, and M.A. Boden, *The Creative Mind. Myths & Mechanisms* (London: Weidenfeld & Nicolson, 1990). For a comment on the bearing of these works on the continuing debate on the nature of invention, see R. Fox, "L'inventeur et l'invention: une approche historienne de le relation entre continuité et discontinuité", *Alliage* (1994), nos. 20–1: 43–53, on pp. 46–8.

interpret nineteenth-century technology in terms of the conflicts and co-existence of savant and analytical forms of work, and to see similar social and cognitive relations in contemporary science and medicine.

Picon's paper, too, spans the Enlightenment and the early nineteenth century, and it identifies an intellectual shift, specifically in what he calls "engineering rationalities", which (like Pickstone's) occurred, or at least was completed, about 1800. The transition, located very precisely in the world of French engineering, was from what Picon describes as a geometrical, Vitruvian, or "classical" rationality to an analytical one, founded on mathematical analysis and calculus. The concept of an engineering rationality in Picon's sense is wide-ranging; it embraces the most deep-seated, even unconscious representations of both nature and society and calls for any study of technological change to extend to change at this profound cultural level as well. Hence the particular shift he describes was manifested equally by a passage from a technical emphasis on order, proportion, and stability to a broader guiding interest in mobility (of goods and people) and fluidity, and by the engineer's new perception of himself as an agent of social and economic as well as purely technical progresss.

Like all theoretical approaches, those advanced in this book are subject to testing and change. They are tools, to be tried in the more empirically grounded work which is the stock in trade of the practising historian. Recently, there has even been some questioning of the value of theory of *any* kind. This sceptical position was debated in 1991 in an exchange in *Technology and Culture* between Angus Buchanan and John Law, with a "Comment" by Philip Scranton.[19] And it was broached again at the Oxford conference in the context of a more wide-ranging workshop on the theme "Does theory matter in the history of techno-logy?", based on papers by Håkon With Andersen, Buchanan, Mikael Hård, and Scranton, with commentaries by Svante Beckman and Gøran Nilssen.[20] One conclusion that might be drawn from the discussions in *Technology and Culture* and at Oxford is that the division between the exponents of narrative and theory is less marked than it appears to be at first sight: it could be argued, for example — as Law did argue in 1991 and as Pinch does again here in *Technological Change* — that even the barest narrative conceals profound historiographical assumptions and so is itself dependent on theory, albeit unspoken theory.

Another, surer conclusion is that we should be wary of universal prescriptions. It will remain the task of the historian to determine which, if any, of the expanding menu of theoretical options, most though not all of them provided by

---

[19] R.A. Buchanan, "Theory and narrative in the history of technology", *Technology and Culture* (1991), **32**: 365–76; J. Law, "Theory and narrative in the history of technology: response", ibid., pp. 377–84; P. Scranton, "Theory and narrative in the history of technology: comment", ibid., pp. 385–93.

[20] The workshop, based on precirculated papers, was organized by Håkon With Andersen and Mikael Hård.

the social sciences, seem most helpful in tackling a particular problem. The choice has to be a pragmatic one, made on the grounds of immediate efficacy, rather than an attempt to show that one approach is always "right" and another "wrong". Picon's insistence that his concept of engineering rationality is an addition to the systems approaches of Gille and Hughes, and not a substitute for them, explicitly recognizes the complementarity of the analytical tools at his disposal. And the more empirical papers in this volume suggest that a flexible eclecticism of this kind is already widely practised. Nevertheless, they do reflect the general recognition, within the discipline, that some kind of theoretical choice, explicit or implicit, has to be made, even if theory is used only as a heuristic scaffolding, to be removed and perhaps rendered invisible once the descriptive structure is in place. Without such a choice, constructing the "thick" description which the contextualization of the central processes of invention, innovation, and transfer demands can all too easily become a vain struggle with an uncontrollable mass of evidence.

The challenge of reconciling "thickness" of description with the clarity and form that a judiciously deployed theoretical model can provide, or help to uncover, is a formidable one with which historians of technology in the 1990s have routinely to grapple. The alternative — of maintaining that all historical contexts are different and that, *ipso facto*, generalization and an engagement with broader historiographical issues are impossible — is as unsatisfactory as a purely theoretical position untried by a consideration of historical evidence. The contributors to *Technological Change* have sought the balance in different ways and with different emphases. In the concluding section on "Technology, politics, and national cultures", for example, MacKenzie's paper shows how sociological insights can be used to inform a finely focussed case study. In the same section, Ian Inkster exploits an explicitly comparative and interdisciplinary methodology in making general points about the nature of technology transfer. His comparison between the attempts to transfer technologies to Japan and Russia from the West in the late nineteenth and early twentieth centuries points, above all, to the often misapprehended difficulties and uncertainties of the process. In Japan the assimilation of new technologies was broadly successful, for reasons that appear to lie mainly in the state's readiness, during the early Meiji period, to undertake institutional changes affecting virtually all aspects of society. The Japanese government's resolve to create a favourable infrastructure in transport and education, and its support for military technology and the systematic acquisition of technical information helped to create an environment in which, from the mid-1880s, private enterprise could profit handsomely from imported technologies. In Russia, by contrast, the early success of the South Russian metalworking and metallurgical project near Ekaterinoslav in the late 1880s and 1890s gave way to failure after 1900, when governmental support diminished (in the face of other, more pressing problems) and a fruitful alliance with foreign financial interests collapsed, leaving the project a prey to the fragility of the economic, social, and political foundations on which it rested.

The importance of political context is equally evident in Yves Cohen's analysis of some of the technological choices that were made in French and Soviet industry in the 1930s. The comparison he explores is between the rigidly centralized Stalinist economic policies that were enshrined in successive five-year plans from 1928 and the freer, "guided" economy of a France traversing a period of recurring political instability. In both countries, the "actors" within industry — whether they were the heads of production plants in the USSR or entrepreneurs and company managers in France — were set on industrial reform. But the results were very different. In the USSR, the integration of manufacturing with politics bred a preoccupation with the engagement of large, obedient workforces and a consequent neglect of R and D, mechanization, and even the organizational adjustments which the labour-intensive Stakhanovite incarnation of Stalinist policy demanded: political priorities, in fact, were impeding industrial advancement, leaving plant managers to make only very limited decisions that affected their immediate workforce. In France, by contrast, industrialists were able to initiate political change in areas (such as trade union legislation) on which their performance depended. To a degree that would have been unthinkable in the USSR, they were masters of their industrial and techno-logical destiny, even in the years of the relatively interventionist and centralist Popular Front between 1936 and 1938.

Cohen's study, like Inkster's, demonstrates the complexity of the circum-stances which can favour or impede the implantation of an imported technology and which are likely to make an analysis rooted solely in one discipline — economic history or the history of technology, for example — inadequate. Morris Low's discussion of a politically and culturally motivated technological nationalism that has grown up in the recent relations between Japan and the USA also offers reflexions on complexity, in a context highly relevant to the 1990s. A growing mutual suspicion, embedded even in what would seem, at first sight, to be the favourable structures of multi-national concerns, is making the transfer of knowledge and techniques in our own day a less straightforward matter. In Japan, the suspicion has fed on a pride in a distinctive national tradi-tion in post-war technology that has found natural support in right-wing political circles, where American evasiveness in sharing technological knowledge, especially in military matters, has caused particular resentment. So could such nationalism go beyond sentiment, to the point of creating yet a further dimen-sion of complexity and a substantial further barrier to transfer? Low's prediction is that the Japanese strategy of establishing major centres for research and development outside Japan and the general growth of international collabora-tion in research will serve, in due course, as a bulwark against this spectre. The strategy also means that any attempt to analyse transfer as a linear process between one country and another is likely to be hopelessly inapplicable to the 1990s; here again, as Low suggests, we seem to be facing a manifestation of post-modernism and the need for an historiographical change to cope with it.

Calls for historiographical changes are not, of course, peculiar to work on the very recent period. At the other end of the chronological range of this volume, and far removed from post-modernism, are two papers that strikingly illustrate the keenness of debate on a period — the Middle Ages — that has risked becoming overshadowed as research on modern and, in particular, twentieth-century technology has become more plentiful. The papers, by Bert Hall and Richard Holt, are inspired by Lynn White's *Medieval Technology and Social Change*, published in 1962 and described by Hall as being now in "respectable middle age". As Hall and Richard Holt show, few books of such modest size have had a comparable effect in the history of technology, and this despite the attacks that were launched on many of White's theses from the start. One of the severest charges that White's critics have made against him is that he was a "technical determinist"; and his conception of the introduction of the stirrup in the eighth century and its subsequent use in warfare as an essential foundation of feudalism has certainly lent plausibilty to the charge. But is this label, with the burden of opprobrium that it carries in current historiography,[21] really applicable to White? Hall thinks not. For him, White was, if anything, a "cultural determinist", someone for whom technology was more likely to be fashioned by culture than to fashion it; an appropriate society, with a supportive culture, was therefore a *sine qua non* if any technology, whatever its virtues, was to be accepted and developed.

Despite Hall's generally sympathetic assessment of *Medieval Technology and Social Change*, he recognizes that White's perception of medieval society as creative in technology and set on the path of technological achievement that was to lead in due course to the great innovations of the Industrial Revolution would have as few supporters now as have most of his arguments about specific artefacts. One reason for this emerges from Holt's paper, which shows how gravely White's view was hindered by the thinness of the evidence which he was able — or which, in certain cases when the evidence was there, he *chose* — to invoke. On one question that Holt treats — White's characteristic claim (following Marc Bloch) that the water mill was essentially a product of medieval ingenuity, rather than a device that was already in widespread use by the beginning of the Middle Ages — the archaeological findings that have been accumulated since White did his work are decisive: excavations in Ireland, where the conditions of preservation are particularly favourable, show that, there at least, the water mill was common by the tenth century. In this case, as in those of the medieval tide mill, the windmill, and the improvements in the quality of church and domestic building, Holt perceives not White's culturally driven wave of unprecedented ingenuity but rather an unremarkable response to precise

---

[21] In an extensive literature on technological determinism, see John M. Staudenmaier's comments, based on a review of contributions to *Technology and Culture*, in his *Technology's Storytellers. Reweaving the Human Fabric* (Cambridge, Mass.: MIT Press, 1985), pp. 134–48.

social and economic trends, such as the growth of population and the invigoration of agriculture and trade, which are well known to have occurred after 1000 AD.

One result of Holt's analysis would be to effect some smoothing of the pattern of technological change in the Middle Ages and to reveal its geographically diverse fine structure. Somewhat similar results have been emerging from another very active area of recent research — the first industrial revolution. Here, cliometric studies have tended to diminish the abrupt, revolutionary character of the economic growth that occurred in Britain between the late seventeenth and mid-nineteenth century. They have shown that the industrial revolution, conceived as what Patrick O'Brien, Trevor Griffiths, and Philip Hunt describe in their paper as a "widely diffused national event", did not come "on stream" until well after 1800. The pattern before then was a much patchier one, with the main advances in technology clustering in a few sectors (notably metallurgy and textiles) and being exploited unevenly in different regions. It was precisely this patchiness which fired the industrial espionage that John Harris discusses in his paper. His evidence too points to the protracted nature of the industrial revolution, viewed as a locus for technological change. As he shows, quite early in the eighteenth century, Britain was already a target for French entrepreneurs wishing to secure skills and techniques that did not exist in their own country. By 1719, the British government felt compelled to enact legislation to impede the emigration of skilled workers; legislation applying to machinery for the woollen and silk industries followed in 1750, and for the rest of the eighteenth century and on into the early nineteenth century similar Acts continued to be passed, with results that Harris shows to have been moderately (though only moderately) successful.

Although the difficulty of securing an effective transfer of workers and innovative machinery from Britain to France was demonstrated repeatedly, the lure of Britain as an object of technological emulation remained strong until the mid-nineteenth century. The perception of the British as a peculiarly inventive people was, as the papers by both Harris and O'Brien, Griffiths, and Hunt show, an enduring one, and it was already widespread before the mid-eighteenth century. O'Brien, Griffiths, and Hunt, however, observe a clear change in the nature of invention in Britain over this long period. Drawing on their prosopographical study of almost 2,500 British inventors, they distinguish between those who were active in the decades separating John Kay and Edmund Cartwright (roughly from the 1730s to the 1790s) and those of the first half of the nineteenth century. With respect to the earlier period, the proportion of inventors coming from the main stream, predominantly Anglican sectors of society and from the non-industrial regions of Britain was too high to sustain either a Weberian interpretation based on the values inherent in the Disssenting tradition or a close correlation between particular inventions or innovations and demand and supply models rooted in economic theory. For work on the nineteenth century, by contrast, when technological change became more a matter of "diffusion, adaptation, and improvement", these sociological and economic models do seem to have some cogency. In that cogency, however, there lies the historiographical snare that leads O'Brien, Griffiths, and Hunt to urge historians of British textile manufacture to take a "long view"; the tempting, easy assumption that explanatory categories that hold good for the nineteenth century can be applied, writ small, to the eighteenth century is, for them, just unsustainable.

There remains one aspect of the processes of invention and innovation to which none of the theories discussed in this volume can be applied. This is the individual inventive act, the study of which has tended to be abandoned by modern historians of technology, who have moved firmly away from heroic treatments of inventors in the Smilesian mould. There is, it must be said, a growing psychological literature on the subject, and it may be that this literature will, in time, stimulate historians to turn again to the exploration of personal creativity, rather than their favoured terrain of contextual study and the analysis of the conditions that have favoured or impeded the recognition, acceptance, and development of a new technology.[22] For the moment, however, the genre of the heroic individual biography remains something of a disregarded curiosity, an historical artefact which has itself to be subjected to analysis.

Christine MacLeod's contribution identifies the genre very precisely, in the extreme form it assumed in England in the third quarter of the nineteenth century, as a product of a campaign for the abolition of the patent system that broke out in the 1850s. The debate that MacLeod describes raised questions of a kind that most historians have had to face in their own work. Even if psychological speculation is rejected, might there nevertheless be grounds for structuring the history of technology around leaps forward construed as possessing at least a large measure of individual achievement? Or does a vision of invention as a response to social and economic circumstances imply that to ascribe an invention or a new process to one person is arbitrary? In the attacks on the patent system that occurred in the 1850s and 1860s, the latter view lay at the heart of the abolitionists' argument, while the case for the defence, mounted most notably by Bennet Woodcroft, rested on a clear belief in the determining role of personal genius. While, as MacLeod observes, the nature of the interaction between the patent controversy and the simultaneous emergence of an heroic conception of the lives of great engineers and inventors is hard to unravel, the work of such writers as Samuel Smiles and Henry Dircks (who saw the capacity for invention as the mark of a special "faculty of the mind") does seem to bear a recognizable stamp of its time. The time, moreover, was brief: by the 1880s, MacLeod argues, inventors might enjoy celebrity in their own life-time but they were most unlikely to enjoy posthumous fame comparable with that which had come to be associated (and is still popularly associated) with a Watt or a Stephenson.

---

[22] For examples of work that reflect the current interest in creativity in technology and in more general aspects of creative thought, see Robert J. Weber and David N. Perkins (eds.), *Inventive Minds. Creativity in Technology* (New York and Oxford: Oxford University Press, 1992), and M.A. Boden (ed.), *Dimensions of Creativity* (Cambridge, Mass.: MIT Press, 1994).

There is no suggestion, of course, that the approaches discussed in this volume exhaust the range of methodological possibilities that are currently being debated and applied. Likewise, the empirical studies are offered as no more than samples of work in progress in a few key areas. It is only too evident that such themes as technological change in Europe in prehistoric times, classical antiquity, and the Renaissance are unrepresented; and the same is true, for example, of the history of domestic technology and of technology in the Third World. The relations between scientific thought and technological practice — a subject on which Donald Cardwell broke rich new ground with his *From Watt to Clausius* almost a quarter of a century ago[23] — have also not been systematically covered. These are all conspicuously active fields of research, although they are not always ones pursued by scholars who would recognize themselves primarily as historians of technology.[24] Indeed, like the papers in *Technogical change*, they illustrate the considerable diversity of the disciplinary settings in which the history of technology is pursued.

This diversity of settings can have the effect of concealing vital strengths. It also constitutes a warning against a disciplinary parochialism that can too easily feed on otherwise laudable moves to professionalize work in the history of technology. Of course, the discipline can only benefit from the creation of formally designated academic posts in the field. But it will also benefit from having its methods and literature known in other disciplines and from recognizing that relevant contributions in those disciplines are just as likely to be good history of technology as a paper published in *Technology and culture* or one of the other specialist journals. The point, once again, is about flexibility and the avoidance of a too prescriptive vision, whether of an appropriate methodology or of a favourable institutional setting.

In assessing the state of the history of technology in the mid-1990s, therefore, it seems important to take a broad view of the possible locations for good work, including locations outside the academic world. Societies and journals can help in this regard, and, by their openness and refusal to recognize a single royal road to quality, most of them are doing so. Flexibility of a different kind is also becoming increasingly evident, though slowly and never painlessly, in the trend towards a more "international" vision of the history of technology — a vision which has fired the International Committee for the History of Technology (ICOHTEC) since its foundation in 1968 and which brought the Society for the History of Technology to Uppsala in 1992 and will bring it to Europe again, to

---

[23] D.S.L. Cardwell, *From Watt to Clausius. The Rise of Thermodynamics in the Early Industrial Age* (London: Heinemann, 1971).

[24] Cardwell, of course, is a notable exception in this respect. From 1963 to 1984, he taught the history of technology at the University of Manchester Institute of Science and Technology, and for many years, as Professor of the History of Science and Technology, held one of only two chairs in Britain that bore any mention of the subject. The other chair, also in the history of science and technology, was that of Rupert Hall at Imperial College of Science and Technology, University of London.

London, in 1996. Here, however, formidable barriers remain, in language and the ease with which we all slip into habits of reading and citation that focus on work in our own academic or cultural tradition. Bertrand Gille seems, to some extent, to have been a victim of this in the English-speaking world, as have other scholars, such as the anthropologist André Leroi-Gourhan, who shared and, in certain respects, inspired Gille's capacious conception of the discipline.[25] Another barrier, less often articulated, is the difficulty of maintaining a reasonable command of literatures from and on more than one country: here, perhaps, long-term collaborative projects and active international networks are the only solution, although even these have their snares in the time and energy which the mechanisms of collaboration and "networking" consume.[26]

In conclusion, I return to my earlier comments about the stages reached by the history of science in the early 1960s and by the history of technology in the 1990s. The comparison between the two cases must give grounds for optimism. The growth, since the Oxford conference of 1961, in the sheer quantity of research in the history of science and the continuing vigour of the attendant methodological debates have been the marks of a discipline that has justly earned its now central and influential place in the world of scholarship in the humanities and the social sciences. It would be extravagant to claim that the history of technology at present is in a comparable position. But the gathering strength of both intellectual and institutional initiatives, of which the "Technological change" conference and this volume are just two examples, suggests that the discipline, especially when broadly defined in the way I have indicated and with due recognition of the variety of contexts in which it can and does flourish, is set on a path at least as promising as that to which historians of science looked forward more than three decades ago.

---

[25] A particularly important work by Leroi-Gourhan that remains untranslated and too little-known among English-speaking historians of technology is his *Milieu et techniques* (Paris: Albin Michel, 1945). There are, of course, equally notable examples in other national literatures. The case of Akos Paulinyi, almost all of whose work has been published in Slovakian or German, is all the more striking since he has had much to say about the industrial revolution in Britain. For a list of Paulinyi's publications, see V. Benad-Wagenhoff (ed.), *Industrialisierung — Begriffe und Prozesse. Festschrift Akos Paulinyi zum 65. Geburtstag* (Stuttgart: Verlag für Geschichte der Naturwissenschaften und der Technik, 1994), pp. 261–4.

[26] The final report on the recently completed Achievement Project contains some pertinent reflexions on an interdisciplinary collaborative project that could provide a model for future initiatives. The Achievement Project was launched in 1990, with the support of the Renaissance Trust, to undertake a study of human achievement, with special reference to patterns of technological innovation and economic growth since 1500. Copies of *The Achievement Project Newsletter* (published at intervals between 1991 and 1995) and other information, including the final report, may be obtained from The Achievement Project Office, 10B Littlegate Street, Oxford OX1 1QT.

# Models

# The Social Construction of Technology: A Review

*Trevor Pinch*

In the last decade "the social construction of technology" has become much in vogue. Not only do a plethora of authors refer to something they call the social construction of technology, but also the approach as a whole is perceived as a school — something that is taken as a challenge to other strands in the history and sociology of technology. In this review I offer some personal reflexions on the development of work in the social construction of technology. The review is intended to be neither systematic nor exhaustive; it is more a personal take on some of the issues.[1]

## VARIETIES OF SOCIAL CONSTRUCTION OF TECHNOLOGY

Where does the term social construction come from, and what does it mean? Its source is to be found in the extremely influential book by Peter Berger and Thomas Luckmann, *The Social Construction of Reality*, published in 1966.[2] Drawing on the phenomenological tradition and particularly the work of Alfred Schutz, Berger and Luckmann observed how the everyday reality of social institutions is actively constructed by ordinary members of society in the course

---

[1] For a much more systematic review, see W.E. Bijker, "Sociohistorical technology studies", in S. Jasanoff, G.E. Markle, J.C. Petersen, and T.J. Pinch (eds.), *Handbook of Science and Technology Studies* (Thousand Oaks, London, New Delhi 1995: Sage, 1995), pp. 229–56.

[2] Peter Bergger and Thomas Luckman, *The Social Construction of Reality. A Treatise in the Sociology of Knowledge* (New York: Doubleday, 1966).

of their mundane social activity. Subsequently, whole areas of scholarship have developed under the slogan "The Social Construction of X", where X stands for some notable institution or aspect of society, such as mental illness, deviance, gender, education, the law, or science. It is the latter school of thought — the social construction of science — which has inspired much of the recent upsurge of interest in the social construction of technology.

In the early 1980s a number of scholars working within the social construc-tivist tradition in the sociology of science turned their attention to technology. If scientific facts — always taken to be the hard case for the sociology of knowledge — could be treated as social constructs, why not technological artefacts too? At the same time a number of historians of technology, notably Edward Constant and Thomas Hughes, were becoming interested in ideas developed in the sociology of science. The marriage of the two groups is commonly heralded to have taken place at a workshop held at the University of Twente in 1982. The subsequent volume from the workshop, *The Social Construction of Technological Systems*,[3] jointly edited by a historian of technology (Thomas P. Hughes), a sociologist of technology (Wiebe E. Bijker), and a sociologist of science (myself), has become something of a flagship volume for the new social construction of technology approach.

Although the sources and precursors for the social construction of techno-logy movement can be readily identified, elucidating what exactly the "social construction of technology" means is much harder. Indeed, the very success of social constructivism has meant that all too often people are ready to label themselves as social constructivists without, it seems, any clear conception of what that doctrine means and entails. Thomas Luckmann, in a piece celebrating twenty-five years of the publication of *The Social Construction of Reality*, re-marked that "whenever someone mentions 'constructivism' or even 'social constructionism' I run for cover these days".[4] I too wanted to run for cover at a recent meeting on explorations in social construction[5] drawn from the fields of education, communications, and social psychology. Social construction in such circles has been given a New Age "touchy feely" twist. Participants talked about becoming empowered through social constructions. One therapist at this meeting, when asked what social construction meant for him, replied, in all seriousness, "My practice has been transformed, I now always add 'perhaps' to any advice I give"! The flyer to a follow-up workshop, on "Experiencing social construction", states:

---

[3] W.E. Bijker, T.P. Hughes, and T.J. Pinch (eds), *The Social Construction of Technological Systems. New Directions in the Sociology and History of Technology* (Cambridge, Mass.: MIT Press, 1987)

[4] Thomas Luckmann, "Social construction and after", *Perspective. The Theory Section Newsletter of the American Sociology Association* (1992), **15**, no. 2: 4–5 on p. 4.

[5] Inquiries in social construction, University of New Hampshire, 3–6 June 1993.

Our forms of life, what we consider to be real and true, are social constructions ... Social constructions enable and energize life as well as limit it .... Living must be viewed as process of ever transforming social constructions... We want ideas to pop our paradigms and expand our socio-cultural sensitivity. We hope for ways of learning about and enhancing the quality of life in the post-modern, socially saturated, technological intense, twenty first century.[6]

Quite!

Even within the narrower confines of the history and sociology of science and technology, social construction can mean a variety of things.[7] Following Sismondo,[8] the most important distinction to draw is between "mild" and "radical" social constructivism. In its mild form, social constructivism is simply equated with science and technology having social components: the science and technology we get has in some sense been influenced or shaped by such social components, whether they be political interests, consumer groups, marketing, gender stereotypes, or whatever. This mild version of social constructivism is to be found in the work of historians of technology such as David Nye in his study of electricity in America,[9] Alex Rowland in his study of the war chariot,[10] and Pamela Mack in her study of the development of NASA's LANDSAT remote viewing satellite system.[11] The technologies they examine are said to be socially constructed in the sense that consumer groups, political interests, and the like all played a role in determining the final form which the technology took.

For example, David Nye argues that every new technology is a social construction and, in his richly documented study, he focusses upon electricity's meanings and uses for ordinary people in America between 1880 and 1940. Nye's work is fascinating because he shows how the technology became woven into the hopes and dreams of users, sometimes in quite unexpected ways. Certainly social reality became constructed in new ways for these users. He also shows convincingly how electricity acquired the predominant meaning in

---

[6] Contribution guidelines to *Experiencing Social Construction. A Handbook of Transformations*, from the Taos Institute, Box 4628, Camino de la Placita, Taos, New Mexico 87571, USA.

[7] See Sergio Sismondo, "Some social constructions", *Social Studies of Science* (1993), **23**: 515–53, for a particularly interesting discussion of different meanings of social construction in the sociology of science and technology.

[8] Sismondo, "Some social constructions".

[9] David Nye, *Electrifying America. Social Meanings of a New Technology, 1880–1940* (Cambridge, Mass.: MIT Press, 1990).

[10] Alex Rowland, "Chariots of the mind: the social construction of the war wagon", Paper delivered to the "Technological change" conference, University of Oxford, 8–11 September 1993.

[11] Pamela Mack, *Viewing the Earth. The Social Construction of the Landsat Satellite System* (Cambridge, Mass.: MIT Press, 1990).

America as a commodity to be consumed by isolated users. But what Nye does not do is to show how the very working of the technology and the different technical options offered to users were socially constructed.

The radical version of social constructivism is concerned to show how social processes influence the very content of technology — what it means for a technology to be deemed as working, for example. This version, which draws heavily upon work in the sociology of science, claims that the meaning of the technology, including facts about its working — established perhaps through a process of engineering design and testing — are themselves social constructs.[12] This latter view is opposed to any conception of technological determinism which posits technology developing under its own immanent logic.[13] There is now a considerable corpus of such work. A good example is Donald MacKenzie's study of the development of ballistic missile technology.[14] MacKenzie shows that in the testing of such missiles there were competing definitions of what missile accuracy meant. Other work within a radical tradition of social constructivism include Wiebe Bijker's study of the development of bakelite and of fluorescent lighting,[15] Pinch and Bijker on the development of the safety bicycle,[16] Elzen on the ultracentrifuge,[17] Thomas Misa on the manufacture of steel,[18] and Paul Rosen on the mountain bike.[19]

In the rest of this review, I intend to focus upon the achievements, problems, and potential of this radical social constructivism. I shall be concerned not only to draw attention to criticisms but also to show how the social constructivist approach can be extended to deal with such criticisms. Before doing so, however, it is worth remarking once more on the confusion between the radical and mild

---

[12] The agenda for the radical view is set out in T.J. Pinch and W.E. Bijker, "The social construction of facts and artefacts: or how the sociology of science and the sociology of technology might benefit each other", *Social Studies of Science* (1984), **14**: 399–441.

[13] Technological determinism is something of a bogey word. For a particularly clear discussion of the issues, see Bruce Bimber, "Karl Marx and the three faces of technological determinism", *Social Studies of Science* (1990), **20**: 333–51.

[14] See, for instance, Donald MacKenzie "From Kwajalein to Amargeddon? Testing and the social construction of missile accuracy", in David Gooding, Trevor Pinch, and Simon Schaffer (eds.), *The Uses of Experiment* (Cambridge: Cambridge University Press, 1989), and Donald MacKenzie, *Inventing accuracy. A Historical Sociology of Ballistic Missile Guidance* (Cambridge, Mass.: MIT Press, 1991).

[15] Wiebe Bijker, "The social construction of bakelite: towards a theory of invention", in Bijker, Hughes, and Pinch, *The Social Construction of Technological Systems*, pp. 159–87.

[16] Pinch and Bijker, "The social construction of facts and artefacts".

[17] Boelie Elzen, "Two ultracentrifuges: a comparative study of the social construction of artefacts", *Social Studies of Science* (1986), **16**: 621–62.

[18] Thomas J. Misa, "Controversy and closure in technological change: constructing 'steel'", in Bijker and Law, *Shaping Technology*, pp. 109–39.

[19] Paul Rosen, "The social construction of mountain bikes: technology and post modernity in the cycle industry", *Social Studies of Science* (1993), **23**: 479–513.

versions of social constructivism, because this confusion has been the source of some recent criticism.

## THE "SO WHAT?" RESPONSE

There is an old adage that new ideas go through three stages of development: First, they are ignored. Secondly, they are explicitly rejected. Lastly, they are accepted, but people say "So what? What's all the fuss about? It's nothing new". In evaluating the progress made by the social construction of technology approach I would say that it is a mark of our achievement that we seem to have reached this last stage. Rather than ignoring the approach, or rejecting it, people now seem to say "So what? Sure we knew technology was a social construct all along — what's new?"

A recent expression of this position has come from David Edgerton in an essay review of MacKenzie's work and other social constructivist studies of technology.[20] Edgerton correctly points out the roots of this approach in the social construction of *science* and argues that, there, the social construction thesis had something novel to say because it was counter-intuitive to challenge the predominant view that science is solely the product of nature. When it comes to technology, however, everyone accepts that it is not exclusively the product of nature; so what's new? Edgerton points out that the modern history of techno logy as exemplified by the work of members of SHOT has long accepted the "embeddedness of technology in the human world".[21] Equating SHOTtish history of technology with the Model T and the SCOTtish (social construction of technology) approach with a Chevrolet, Edgerton asks:[22]

> Why go for the annually restyled Chevrolet when the Model T will do just as well?[23]

It is clear that Edgerton's criticisms here are directed towards the mild form of social constructivism. And perhaps he is right. The mild form of social

---

[20] David Edgerton, "Tilting at paper tigers", *The British Journal for the History of Science* (1993), **26**: 67–75.

[21] Edgerton, "Tilting at paper tigers", p. 74.

[22] Edgerton, "Tilting at paper tigers", p. 75.

[23] Unknown to Edgerton, in collaboration with the Cornell historian of technology, Ron Kline, I have been examining the early history of the motor car, including the Model T. We have found that this paradigmatic "black box" was actually opened for a while by rural consumers, who found a radically different meaning in the Model T. In understanding this development, a SCOTtish approach turns out to be rather useful. Perhaps, just as the actual artefact, the Model T has not been as stable as is assumed in conventional historical accounts, SHOT itself may turn out to be less monolithic than some historians have thought. See Ron Kline and Trevor Pinch, "Taking the black box off its wheels: the social construction of the American rural car", submitted to *Technology and Culture*.

constructivism may well simply be the latest accessory to be added to the Model-T. Mild social constructivism allows historians to repackage what they have always done in a new way. But the stronger form of social constructivism does offer something more. It points to technology being through and through social; the implications, as we shall see, can be quite radical.[24]

Before going further, it is necessary to spell out even more precisely what the radical social constructivists' agenda is.


## THE SOCIAL CONSTRUCTION OF TECHNOLOGY: THE RADICAL AGENDA

The work that came out of the melting pot of sociology of science and history of technology in the early 1980s has led to three broadly distinguishable but over-

---

[24] The attempts by recent commentators, such as Angus Buchanan and David Edgerton, to drive a wedge between the fields of sociology of technology and history of technology belie the *de facto* links which gave birth to SCOT and which continue to enrich it.

An illustration, drawn from personal experience, of the salience of the links between sociology and history, is as follows. The conference and subsequent volume *The Social Construction of Technological Systems* (see above, note 3) is often taken as the birthplace of the social construction of technology approach. This conference and publication involved sociologists of science and technology and mainstream American historians of technology, such as Thomas Hughes, Edward Constant, and Bernie Carlson. I am sure that no forced marriage was intended; neither was the child that was born (SCOT) a product of immaculate conception. Far from it. Those of us who were involved in planning the workshop and editing the volume just took it for granted that to approach the problem of adequately understanding technological change required both the special competence and skills of sociologists and of historians (plus the practitioners of many more diciplines). In other words, and to stretch the obstetrics metaphor to breaking point, the social construction of technology came from natural child birth. That seemed at the time the best way to proceed, and for me it continues to be the best way to do so. As long as historians and sociologists address common problems and see the merits of one another's approaches, they will work together. Now that SCOT is out in the world, I continue to find that the history of technology is the natural place to turn to in doing this sort of work. In my own recent research on the rural car, for example, I have found it perfectly appropriate to collaborate with Ron Kline. Both of us found that the sociological and historical dimensions are indispensible to our task. It is not (borrowing David Edgerton's terms) that SCOT-work and SHOT-work are different things. They naturally merge together.

lapping models of technology: the social construction of technology (SCOT),[25] actor-network theory,[26] and the systems model.[27]

What these approaches had in common was their attempt to understand how a variety of social, political, and economic considerations shape technological development. The one pervasive metaphor that encapsulates all three approaches is that of the "seamless web".[28] Technology forms part of a seamless web of society, politics, and economics. Thus, the development of a technological artefact, such as a high-resistance incandescent lamp, is not merely a technical achievement; embedded within it are societal, political, and economic considerations. All three approaches are concerned, to varying extents, to address the social and technical in equivalent ways. The hardest part in any such analysis, of course, is to show *how* the very artefacts themselves contain society embedded within them. "Opening up the black box of technology" became the rallying cry for the new work.

Here, I deal mainly with what has become known as the SCOT approach. This is because the SCOT approach, as its name designates, is the most explicitly social constructivist of the three, and, for reasons of clarity, it is best to address the issues within the confines of that approach. Many of the criticisms I shall mention, however, apply equally to all three approaches. Also it is worth pointing out that authors in their studies have combined features of the different approaches.[29]

What, then, were the essentials of the SCOT approach as it was first formulated? In any attempt to understand technological development, it is axiomatic that the technological artefacts (broadly construed to include materials and processes) play a prominent part in the analysis. SCOT, being fundamentally a sociological approach towards technology, analyses artefacts in the context of society. The particular way in which society is conceptualized and linked to artefacts is via the notion of *relevant social groups*. These are identifiable social

---

[25] See Pinch and Bijker, "The social construction of facts and artefacts".

[26] See Michel Callon, "Society in the making: the study of technology as a tool for sociological analysis", in Bijker, Hughes and Pinch. *The Social Construction of Technological Systems*, pp. 83–103; John Law, "Technology and heterogeneous engineering: the case of the Portuguese expansion", ibid., pp. 111–34: and Bruno Latour, *Science in Action. How To Follow Scientists and Engineers Through Society* (Milton Keynes: Open University Press, 1987).

[27] Thomas P. Hughes, *Networks of Power. Electrification in Western Society, 1880–1930* (Baltimore: Johns Hopkins University Press, 1983), and Thomas P. Hughes, "The evolution of large technological systems, in Bijker, Hughes, and Pinch, *The Social Construction of Technological Systems*, pp. 51–82.

[28] T.P Hughes, "'The seamless web': technology, science, etcetera, etcetera", *Social Studies of Science* (1986), **16**: 281–92.

[29] For example, MacKenzie has used aspects of the systems approach and the actor network approach in his *Inventing Accuracy*. Elzen, in "Two ultracentrifuges", tries to combine elements of the actor-network approach with SCOT.

groups that play a role in the development of a technological artefact. The key element is that such groups share a meaning of the artefact — a meaning which can then be used to explain particular developmental paths which the artefact takes. There are many possible groups, including engineers, advertisers, public-interest groups, consumers, and so on. When dealing with a complex technology, a whole array of such groups may be identified. Although the only defining property is a homogeneous meaning given to a certain artefact, the intention is not just to retreat to well worn general statements about "consumers" and "producers". A detailed description of the relevant social groups is needed in order better to define the functioning of the artefact with respect to each group.

In the case of the development of the bicycle, Pinch and Bijker focussed upon two particular groups.[30] These were: (1) "young men of means and nerve: they might be professional men, clerks, schoolmasters, or dons", they shared a meaning of the high wheeler as a virile machine for travelling fast, for showing off to one another, and for sport rather than transport; and (2) "women and elderly men" who wanted a bike for transport and who shared a meaning of the high-wheeler as the "unsafe machine". Each of these groups had different problems with the then current high-wheeler machines. For example, elderly men had the "safety problem", whereby the rough roads would lead them to be thrown from the bicycle (called doing a "header"). Young men had rather the opposite problem; they wanted to use the bicycle for racing and, for those purposes, they needed it to go even faster. Different solutions to these problems are apparent. For example, one solution to the safety problem was to slope back the front fork. This led to artefacts, such as the Singer Extraordinary bicycle of 1878. One solution to the speed problem was to make the front wheel even larger; this led to the Rudge Ordinary of 1892, with its enormous 56-inch wheel. The bigger the wheel, the faster, the more dangerous, and the more desirable this "macho bicycle" became for the relevant social group.

Following the developmental process in this way, it was possible to see growing and diminishing degrees of stabilization of different artefacts. By using the concept of stabilization or "closure", the invention of the safety bicycle was treated not as an isolated event, but as a nineteen-year process with a variety of different artefacts achieving different degrees of stabilization for different social groups.

In examining technological development in this way Pinch and Bijker found it useful to take over the notion of "interpretative flexibility" from the sociology of science. By this it is meant that radically different meanings of an artefact can be identified for different social groups; there is interpretative flexibility over the meaning to be given to the artefact. The high-wheeler had the meaning of the "macho machine" for young men of means and nerve, but for older people and women it had the radically different meaning of the "unsafe machine". Such interpretative flexibility may apply not only to a compound artefact but also to

---

[30] Pinch and Bijker, "The social construction of facts and artefacts".

some component of it. For example, when the air-tyre was first introduced, it was for some groups an object of derision, aesthetically unappealing, and a source of endless trouble (punctures). On the other hand, for Dunlop it was the perfect solution to the problem posed by the vibrations of the bicycle. (Remember that most bicycle riding took place on very rough roads indeed.)

Interpretative flexibility means that an artefact can be given radically different meanings, such as in the case of the high-wheeler discussed above. The different meanings may then become embodied in new artefacts. It is unlikely, however, that this interpretative flexibility will continue for ever. What one observes is that closure and stabilization occur in such a way that some artefacts appear to have fewer problems and become increasingly the dominant form of the technology. This, it should be noted, may not lead to all rivals vanishing, and often two very different technologies may exist side by side (for example, jet planes and propeller planes). Also this process of closure or stabilization need not be final. New problems can emerge and interpretative flexibility of the artefact can once more appear. In describing the process of closure and stabilization, Pinch and Bijker found it useful to talk about particular "closure mechanisms" (again an idea borrowed from the sociology of science).

The adaptation of the air-tyre was explained by its use in bicycle races, these races served to close down debate over the merits of the air-tyre. Dunlop's air-tyre, as mentioned above, was for most engineers a theoretical and practical monstrosity. For the general public as well, it was aesthetically unappealing. When, for the first time, the tyre was used at the racing track, its entry was hailed with derisive laughter. However, this was quickly silenced by the high speed achieved, and there was astonishment when bicycles with air-tyres outpaced all rivals. With respect to two important groups, the sporting cyclists and the general public, closure had been reached, but not by convincing those two groups of the feasibility of the air-tyre as an anti-vibration device. This "redefinition of the problem", from vibration to speed, is an example of what Pinch and Bijker called a "closure mechanism".

The importance of races in bringing about closure for a variety of audiences is apparent throughout the history of the bicycle. Closure means closure for some specific social group, and races (often reported upon in the press) are one place where many different social groups can be reached. Likewise, races have been important in the history of the car. The pneumatic car tyre also only became a stabilized artefact after its successful use in car races.

There is no doubt that SCOT, as originally conceived, offered the bare bones of an approach for doing social studies of technology. Over the years, a number of authors have built upon SCOT,[31] refining the approach so that even if it is not

---

[31] See, for example, Hughie Mackay and Gareth Gillespie, "Extending the social shaping of technology: ideology and appropriation", *Social Studies of Science* (1992), **22**: 685–716; Elzen, "Two ultracentrifuges"; and Rosen, "Social construction of mountain bikes".

quite the mountain bike of technology studies (that perhaps is yet to come), it has proved to be quite durable — a good solid "roadster". Perhaps the most important development in SCOT has come from Bijker himself, who has added the important notion of a "technological frame".[32] Bijker's frame is like a "frame of meaning" associated with a particular technology which is shared between several social groups and which further guides and shapes the development of artefacts. With this concept, Bijker has been able to provide a link between the wider society in which the technology is immersed and its developmental path — something that was left uncomfortably unspecific in the original SCOT model.

So what was radical about SCOT, and why is it simply not, as Edgerton and other critics assert, the old human embeddedness of technology thesis? It is radical, I suggest, precisely because of the crucial notion of "interpretative flexibility" of artefacts. What is being suggested is that an artefact, including its workability, can be subject to radically different interpretations which are co-extensive with social groups. This is not the same as saying simply that technology is embedded in human affairs. SCOT focusses attention upon what counts as a viable working artefact and what indeed counts as a satisfactory test of that artefact. In the area of technological testing, for example, SCOT has much to offer. I have already mentioned MacKenzie's study of the testing of ballistic missiles. Another study is my own work on the testing of clinical budgeting systems in the British National Health Service.[33] Such studies show that crucial similarity judgements have to be made in projecting any set of results from the circumscribed environment of the test to the actual use of the technology. In practice, such judgements can be challenged, with critics asserting that the test environment is different from the environment of use. One dramatic case where such judgments actually became crucial was on the eve of the space shuttle Challenger disaster.[34] At a teleconference preceding the fatal launch decision, some critics argued that O-ring test data could not be extrapolated to the low temperatures of launch expected later that night. NASA managers, on the other hand, argued that such an extrapolation was perfectly valid. In short, there was interpretative flexibility over this set of test data. The testing of a technology and whether it is deemed working or not is an inescapably social process.

To make the point another way: it would be unsatisfactory if a social constructivist analysis of, say, a satellite system were simply based upon showing that the final hardware and what it was supposed to do depended upon a set of negotiations amongst a complex web of agencies and different user groups. A more satisfactory analysis would have to show how the actual technical working

---

[32]  See Bijker, "Social construction of bakelite".

[33] Trevor Pinch, "'Testing, one, two, three ... testing!' Towards a sociology of testing", *Science, Technology and Human Values* (1993), **18**: 25–41.

[34] Ibid.

of the system was itself embedded in social choices and negotiations. To make the point in yet another way, in terms of the sociology of science: it would not be satisfactory to make a claim that high energy physics is socially constructed by simply pointing to the political negotiations over the super-collider. Certainly those negotiations will, to some extent, determine whether a collider gets built and what form it may take, but this does not show that the particles detected at the collider, which make up the content of high energy physics, are products of political or social negotiation. The type of analysis carried out in the recent sociology of science has shown how the very entities of modern physics are socially constructed.[35] It is such approaches that SCOT attempts to emulate.

## OTHER CRITICISMS OF SCOT

In addition to the "so what?" response, a number of other criticisms have been levelled at the SCOT approach over the years. Here are some of the main ones:

(1) The jargon of social constructivism is obscurantist; nothing is added which good plain narrative history cannot do better.[36]
(2) The approach is too formulaic.[37]
(3) There is too narrow an emphasis upon the design stage and the early development of a technology, without sufficient attention being paid to the users of technology.[38]
(4) The social relationships and power structures between different social groups and the macro-political context in general are ignored or under-played.[39]

---

[35] See, for example, Andrew Pickering, *Constructing Quarks* (Chicago: University of Chicago Press, 1985), and Trevor Pinch, *Confronting Nature. The Sociology of Solar Neutrino Detection* (Dordrecht: Kluwer, 1986).

[36] Angus Buchanan, "Theory and narrative in the history of technology", *Technology and Culture* (1991), **32**: 365–76, and Angus Buchanan, "The poverty of theory in the history of technology", paper presented to the "Technological change" conference, University of Oxford, 8–11 September 1993.

[37] Langdon Winner, "Upon opening the black box and finding it empty: social constructivism and the philosophy of technology", *Science, Technology and Human Values* (1993) **18**: 362–78, and Steve Woolgar, "The turn to technology in social studies of science", *Science, Technology and Human Values* (1991), **16**: 20–50.

[38] See MacKay and Gillespie, "Expanding the social shaping of technology", and Stewart Russell, "The social construction of artefacts: a response to Pinch and Bijker", *Social Studies of Science* (1986), **16**: 331–46.

[39] See Russell, "Social construction of artefacts".

(5) Social constructivism is politically insipid, and although it provides a successful academic analysis, it avoids the real politics of artefacts.[40]

Of course, as one would expect for any fairly well developed research
agenda, numerous other questions have been raised.[41] However, I think the
above list captures the main points of contention. I shall briefly examine each of
the above criticisms in turn. I shall also point to additions to SCOT, made by
Kline and myself[42] and by Bijker,[43] which try to address some of these
criticisms. Like any dynamic research programme, SCOT has been modified
somewhat over the years to adapt it to the new groups of users in science and
technology studies.

## (1)   SCOT is jargon-ridden — narrative history is better

This criticism has been levelled by Angus Buchanan.[44] I think that it can easily
be dealt with, as John Law has already responded to Buchanan on this point in a
debate in *Technology and Culture*.[45] As Law points out, narrative history is certainly not itself neutral and has its own embedded theoretical assumptions about
the nature of technology and about the nature of history. SCOT is perhaps more
forthright in spelling out the types of assumptions it makes. Integrating SCOT's
conceptual analysis with empirical case studies remains one of the most difficult
things of all to do, and certainly many of the articles published in the SCOT
*oeuvre* still suffer from a lack of integration. The task is hard, but I believe that
the best analyses, such as MacKenzie's recent book,[46] show convincingly how
concepts and empirical study mesh together. Whether narrative history is found

---

[40] Winner, "Upon opening the black box".

[41] A recurrent issue is how the term "social group" can be operationalized. The response
that Bijker and I always give is that once an empirical study is embarked upon, historians
and sociologists have little trouble, in practice, in identifying the relevant social groups and
their boundedness. A related issue is how to identify social groups that are marginalized.
There is nothing in SCOT which says that social groups must arise from the empirical
material alone; obviously groups that one may think are important for a *priori* reasons can
be searched for. Sensitivity must be exercised in looking at unusual sources to get at the
marginalized groups.

[42] Kline and Pinch, "Taking the black box off its wheels".

[43] Weibe Bijker, *On Bikes, Bulbs and Bakelite. The Social Construction of Technology*
(Cambridge, Mass.: MIT Press, 1994)

[44] Buchanan, "Theory and narrative", and Buchanan, "Poverty of theory".

[45] John Law, "Theory and narrative in the history of technology: response", *Technology and
Culture* (1991), **32**: 377–84.

[46] MacKenzie, *Inventing Accuracy*.

to offer more than SCOT will depend upon whether readers judge that SCOT studies have thrown fresh light upon the development of technology.

As to the question of jargon, SCOT like most social science approaches has its fair share of analytical terms. In my view, however, such terms can clearly be explained. Most of us in our writing have tried to explicate in as plain a way as possible what such terms as "interpretative flexibility", "closure", and so on mean. If we have not been clear enough, we must try harder. Obviously, as the aim is to break from common sense descriptions of technology, some specialist language and concepts will be inevitable.[47]

## (2)   SCOT is too formulaic

In a way, this is exactly the opposite criticism to that just discussed. Rather than SCOT being opaque and obscurantist, this line of criticism suggests that the approach is so clear and easy to apply that it can be used in a formulaic manner. SCOT offers a particular language and method for understanding technology. Unfortunately some people may have taken the principles to be cast in stone. Langdon Winner writes:

> It [SCOT] offers clear, step-by-step guidance for doing case studies of technological innovation. One can present this method to graduate students, especially those less imaginative graduate students who need a rigid conceptual framework to get started, and expect them to come up with empirical studies of how particular technologies are 'socially constructed'.[48]

If the principles of SCOT are applied mechanistically, then the result will invariably be disappointing. I do not know which "less imaginative graduate students" Winner has in mind, but most of the work I am aware of in the SCOT genre usually ends up modifying or adding to the approach rather than slavishly following it.[49] On the other hand, not to offer any analytical tools for studying technology can result in naively unsatisfactory stories. If we are to advance the field and build upon case studies, some sort of uniformity in analytical approach may not be such a bad thing. Probably all successful approaches suffer from accusations of becoming formulaic. As I have remarked in my detailed reply to

---

[47] I might add that my own experience of teaching SCOT for several years to a variety of students drawn from the social science and engineering is that they find little difficulty in understanding the key ideas and texts. Indeed, the Open University uses Pinch and Bijker, "The social construction of facts and artefacts" as a teaching text.

[48] Winner, "Upon opening the black box", p. 366.

[49] See for example, Elzen, "Two ultracentrifuges", and Rosen, "Social construction of mountain bikes".

Steve Woolgar,[50] who also accuses SCOT of being formulaic,[51] even his reflexive critique is in danger of becoming formulaic.

## (3)   SCOT focusses upon the design stage of technology at the expense of users

SCOT, as originally conceived, dealt mainly with the design stage of a technology. Interpretative flexibility took the form of identifying different design options, only some of which were stabilized. Perhaps the notion of closure in the sense that one design became dominant was a little too rigid. What was missing was a sense of how and in what circumstances a technology could be modified, particularly as it was taken up and used by different social groups.

I think this criticism of SCOT is essentially correct, and Kline and I[52] have recently tried to address the objection in the particular case of the development of the automobile in rural America. Bijker[53] too, in his analysis of the fluorescent lamp, moves away from the design stage of technology to consider social construction in the process of diffusion.

To give a flavour of the on-going research in SCOT, I shall elaborate upon our work on the car. Our research has shown that in the rural context farmers found a meaning for the car radically different from its meaning as a form of transport. Perhaps the most dramatic new means is that shown in the photograph in Fig. 1. Here a Model T is jacked up in the farmyard to provide a standing source of power to operate a washing machine. A variety of agricultural machines, and less often domestic machines, were powered in this way. Kline and I discuss such developments in terms of interpretative flexibility over the car reappearing, but, unlike in the original SCOT model, the flexibility is not at the design stage. Radically new and different meanings are being given to the car by the social group of users, in this case farm men. Rather than the car being used solely as a source of transport, a new meaning is being added in terms of the car being used as a standing source of power. Further uses soon appeared, such as for the transportation of hay and for ploughing.

The importance of drawing attention to such interpretative flexibility is that it provides a key to historical questions that have to be investigated. What happened to such different meanings given to the car? Did they vanish, and if so, why? Our research suggests that closure over the meaning of the car was reestablished by 1945, by which time the radically different meanings of the car had all but vanished. But all sorts of interesting events unfolded before this closure was achieved. Kits were manufactured (but not by the car

---

[50] Trevor J. Pinch, "Turn, turn and turn again: the Woolgar formula", *Science, Technology and Human Values* (1993), **18**: 511–22.

[51] Woolgar, "The turn to technology".

[52] Kline and Pinch,"Taking the black box off its wheels".

[53] Bijker, "The social construction of flourescent lighting".

**Figure 1**   A jacked-up Model T on a Kansas Farm. I am grateful to Ron Kline and his family for permission to use this photograph.

manufacturers) to facilitate the use of the car as a standing source of power and as a tractor. Meanwhile car manufactures, such as Ford, discouraged using the car in this form and developed dedicated machines for some of the new uses — for instance, as tractors for ploughing. Also, with electrification, standing electric power sources became more widely available.

I cannot go into detail here, but basically it seems that SCOT can be extended to include the use of technology; when this is done, elementary SCOT concepts such as interpretative flexibility, closure, and social groups are found to be fully applicable.

## (4)   SCOT ignores power relationships

As many commentators have noticed, SCOT seems to have had little to say about the social structure and power relationships within which technological developments take place. Pinch and Bijker[54] have defended this lack of attention by pointing to the strategic importance of reorientating technology studies back towards the artefacts and away from social theory. They argued that there was more than enough social theory to go around for everyone but not enough detailed understanding of specific artefacts and of the role society played in shaping them. This reorientation is no longer so crucial; we now have plenty of case studies of how artefacts and society are entwined together. There is nothing, in principle, that prevents a SCOT approach from considering power structures

---

[54] Trevor J. Pinch and Wiebe Bijker, "Science, relativism and the new sociology of technology: reply to Russell", *Social Studies of Science* (1986), **16**: 347–60.

and social relationships between social groups. Indeed, in his recent work Bijker has explicitly dealt with such issues of power.[55]

A related criticism is SCOT's neglect of issues of gender,[56] even though, as Cynthia Cockburn and Susan Omroyd have recently pointed out,[57] Pinch and Bijker in their work on the bicycle did at least draw attention to the importance of considering the social group of women bicyclists. Kline and I have since taken up this gender issue specifically for the case of the rural car. We argue that the remarkable interpretative flexibility of the rural car has a strong tie to gender relations between farm men and women. The overall patriarchical structure of the rural farmstead and the gendered division of labour meant that in general the men performed what are traditionally regarded as the main income-producing activities, in the field, barn, and machine-shop; the women performed "supportive" tasks (from the men's point of view) in the house, garden, and poultry shed. Within this patriarchal structure, which was more flexible than is generally assumed, there were gender identities among farm men and women that help to explain the social construction of the rural automobile. Farm men, especially in the Mid West, saw themselves as technical virtuosi who could operate, maintain, repair, and redesign any and all technologies on the farm. The technical competence of farm men, as with other male social groups, was a core element of their masculine gender identity. Women might pump water, drive the horse and buggy to town, and occasionally operate field machinery, but men fixed a leaky pump, oiled and greased the buggy, and redesigned a hay binder to work over hilly ground. Technical competence helped to define them as masculine (in opposition to a non-technically competent femininity) and reinforced the patriarchal system.

Consequently, the gasoline automobile, which was already steeped in masculine symbolism,[58] came onto farmsteads where men held sway, in part, because of their technical competence. The early car was usually viewed as the latest, highly sophisticated piece of farm machinery, and it became, by virtue of this, the province of men. Women drove the car, as they had the horse and buggy, but men maintained, repaired, and endlessly tinkered with the new addition to the farmstead. The farm men's strong sense of masculine identity, based on technical competence, thus enabled them to re-open the black box of the car, jack up its rear wheels, and power all kinds of men's work on the farm; less frequently, it would also power the woman's cream separator, water pump, or washing

---

[55] Bijker, *On Bikes, Bulbs and Bakelite.*

[56] See Judy Wajcman, *Feminism Confronts Technology* (University Park: Penn State University Press, 1990).

[57] Cynthia Cockburn and Susan Omroyd, *She, He and It. Gender and Technology in the Making* (London: Sage, 1994).

[58] See Virginia Scharff, *Taking the Wheel. Women and the Coming of the Motor Age* (New York: Free Press, 1991).

machine. The gendered division of labour could not be more striking than in the photograph shown in Figure 1, where the man has jacked up the car but the woman still operates the washing machine.

The mutual interactions between the artefact, social groups, and inter-group power relations are clearly evident in this case. The gender identity of farm men, formed by defining it in contrast to the femininity of farm women, enabled men to interpret the car flexibly and to socially construct it as a stationary power source. This social construction, in turn, helped further to shape farm men as technically competent, i.e. as masculine, so reinforcing their gender identity vis-a-vis farm women.

I have gone into this example in some detail, not only because it is one with which I am very familiar, but also to show how SCOT can throw light on issues to do with social structure and power relationships. So those who criticize SCOT for its failure to treat these issues have a point. But this is not an inherent limitation in the approach, rather it reflects the early work within SCOT, which was located at the design end of technology and tended to avoid looking at the larger constellation of factors that shaped technology.

## (5) SCOT is politically insipid

One of the most vigorous attacks on SCOT has come in a recent piece by Langdon Winner provocatively entitled "Upon opening the black box and finding it empty".[59] Winner's main criticism of SCOT is a political one. He writes:

> Unlike the enquiries of previous generations of critical social thinkers, social constructivism provides no solid, systematic standpoint or core of moral concerns from which to criticize or oppose any particular patterns of technical development. Neither does it show any desire to move beyond elaborate descriptions, interpretations and explanations to discuss what ought to be done.[60]

Elsewhere, he writes:

> The attitude of social constructivists seems to be that it is enough to provide clearer, well-nuanced explanations of technological development. As compared to any of the major philosophical discussions of technology, there is something very important missing here; namely, a general position on the social and technological patterns under study.[61]

Winner is surely right. People working within the SCOT approach tend not to make over-arching pronouncements for or against the particular technological

---

[59] Winner, "Upon opening the black box".

[60] Ibid., p. 374.

[61] Ibid., p. 375.

developments they are studying. They fail to do so, I suggest, not because of a lack of moral fibre or political will-power but because they know enough to realize the complexities they are examining and the futility of trying to change the world by pronouncements. What Winner sees as a disadvantage — the paucity of pronouncements for or against particular technologies — is actually one of the strengths of SCOT. What Winner also fails to understand is the radicalness of social constructivism. The old safe dichotomies of critical theory and Marxism of society, social interests, and politics, on the one side, and technology, artefacts, and machines, on the other, are being dissolved. We live in a much more interesting world, one where machines and people are in much closer interaction and where users, such as the rural farming community, may socially construct a machine in a wholly unexpected way. This is not to say, however, that SCOT should not have a political dimension. How could it avoid having one?

So while Winner is correct in calling attention to the neglect of politics of technology in the SCOT approach, I doubt whether the moral certainty that he calls for is easily obtainable. The earlier generations of critical thinkers about technology, represented by Jaques Ellul, for example, were able to make their pronouncements because they adhered to a version of technological determinism. Technology could be pointed to as the evil variable, which in some sense corrupted or repressed human kind. But if it is our society which is embedded in machines, then such simple and sweeping generalizations about the evil of technology are called into question. Winner, in his own work, makes distinctions between different ways in which technologies can be political. But, in condemning particular technologies, such as nuclear power, because of the repressive state needed to operate such technologies, he is forced once more into a version of technological determinism.[62] That is to say, for Winner the social relations that surround particular sets of technological developments inevitably have to take a particular character or form. Because of fears of terrorism or the consequences of plutonium falling into the wrong hands, the technology determines a particular sort of social relations. What Winner fails to examine is whether the technology of nuclear power has to take the form that it does. The SCOT notion of interpretative flexibility draws attention to seeking alternatives and in that way can serve as a corrective to over-pessimistic views of technological development.

What the social constructivist work points to is that the design and adaption of technology should be part of the political agenda. In other words, these issues should be opened up for debate among wider constituencies than at present. There is no one inevitable logic of development. There is choice. And this draws attention to the technology we never get. Here, I think feminist work on

---

[62] Langdon Winner, "Do artifacts have politics?", in *The Whale and the Reactor* (Chicago: University of Chicago Press, 1986).

technology (such as that on domestic technology by Ruth Schwartz Cowan[63] and Dolores Hayden[64]) has been very important in pointing to alternative designs that question the often implicit assumptions about, say, how household labour is to be organized.

If SCOT and other approaches do indeed obtain a better understanding of how technical change occurs, it is hardly imaginable that there are not implications for both politics and policy.

# CONCLUSION

In this brief review, I have only been able to highlight some of the lively debate that SCOT has generated and is generating. When Wiebe Bijker and I first talked of social construction of technology, as a way of understanding better why some bicycles like the "Invincible" (a form of the Rudge Ordinary bicycle) had such big wheels, we had no idea that ten years later we should be accused (as we have been by Winner) of being "self-consciously imperial".[65] Certainly, we think we are onto something with SCOT, and we welcome the occasion to persuade others that this is so. It is a paranoid myth, however, that we have vast programmes of research under way in SCOT and are actively prosletysing. Most of us are too busy doing other things. Indeed, I have never yet trained a single graduate student to do research in the SCOT approach, not even a "less imaginative" one!

Why, then, do some historians and philosophers feel so threatened by our studies and approach? Winner suggests that SCOT's desire to go beyond previous work in the philosophy of technology has something to do with an Oedipus complex. I suggest that the explanation is rather more straightforward. It is simply the case that the combination of detailed empirical research with growing theoretical sophistication about science and technology offers genuine new insights into technical change. That is where all the energy and enthusiasm come from. Perhaps just as technical change can lead to resistance, it is not surprising that change in our understanding of it meets a similar reaction. The black box of SCOT is still not yet sealed. It is when a field is still an open box that it is at its most exciting.

---

[63] Ruth Schwartz Cowan, *More Work for Mother. The Ironies of Household Technology from the Open Hearth to the Microwave* (New York: Basic Books, 1983).

[64] Dolores Hayden, *The Great Domestic Revolution* (Cambridge, Mass.: MIT Press, 1982).

[65] Winner, "Upon opening the black box."

# Towards a History of Technological Thought

*Antoine Picon*

## TECHNOLOGICAL SYSTEMS AND TECHNOLOGICAL THOUGHT

To study the broad canvas of technological change, it is necessary to understand what exactly is changing and to characterize the general equilibria that massive innovations upset. Technological change cannot be described simply as a succession of controversies and minute adaptations, of the kind that most work in the sociology of knowledge treats. It is also the result of global displacements that can best be interpreted with the help of the notion of technological system. Historians have proposed various definitions of this notion. But two definitions seem to me especially significant. They are complementary in their approach both to patterns of routine technological activity and to technological change seen as a system shift.

The French historian Bertrand Gille gave the first of my two definitions in his general survey of the history of technology. Struck by the fact that "within certain limits and as a very general rule, all techniques are, to varying degrees, dependent on one another, so that there is necessarily some coherence between them",[1] Gille tried to define and describe this coherence. He began by considering elementary structures corresponding to what he called "a single technical act", such as the use of a tool with the manual procedures (or "gestes") specifically associated with it, or the utilization of an individual machine designed for a particular limited

---

[1] B. Gille, "Prolégomènes à une histoire des techniques", *Histoire des techniques* (Paris: Gallimard, 1978), pp. 1–118, on p. 19. The translation is the author's own. Cf. the slightly different wording used in the English translation of Gille's book: *The History of Techniques* (2 vols., New York: Gordon and Breach, 1986), vol. 1, p. 17.

purpose. At a different level, he introduced technological sets as combinations of these elementary structures in ways that would achieve larger goals, such as the production of iron in a blast furnace. Those sets were, in their turn, part of broader entities, called by Gille technological "paths" — in French "filières techniques"; an example would be the whole sequence of processes leading from the mine to the blast furnace. On this analysis, a technological system could then be defined as "the coherence, at different levels, of all the technical structures, of all the technological sets and paths"[2] that coexist at any particular time. Looking back at the historical development of technology and using his definition of technological system, Gille made — in accordance with this definition — a crucial distinction between the age of "classical systems", centred on the use of water and wood, and the first industrial system, based on steam, coal, and iron. On this perspective, periods of major technological change, such as the first industrial revolution of the late eighteenth century and the first half of the nineteenth century, appeared as transitions leading from one system to another.

French historians have drawn heavily on Gille's approach. But although his conceptual frame is stimulating, it is difficult to relate to more socially oriented studies of technology and technological change. In filling the gap between Gille's systemic conception and the precise analyses of individual institutions and professions, Thomas P. Hughes's definition of technological systems provides important additional tools.[3] Hughes's definition is meant to establish a strong link between a technology and the institutional and professional organizations that create and sustain it, and it has enabled him to describe the development of electrical power in the Western world in a very convincing way.[4] Since Hughes's work is well known among Anglo-Saxon historians of technology, I shall not present it in detail. But, less global as they are than those of Gille, Hughes's systems appear less narrowly constructed, since they include such heterogeneous elements as human organizations and technical artefacts. However, what they lack in conceptual precision is amply counterbalanced by their empirical fecundity.

The approaches of Gille and Hughes represent major contributions to the history of technology. In this paper, I should like to propose an addition to these two systems-based approaches by focussing on technological thought and its evolution, especially the collective mental frames within which actors of production, such as managers, engineers, or workers, think and act. As I shall try to show, these mental frames characterize types of knowledge and reasoning as

---

[2] B. Gille, "Prolégomènes", p. 19; English edition, vol. 1, p. 17.

[3] T.P. Hughes, "The evolution of large technological systems", in W.E. Bijker, T.P. Hughes, and T.J. Pinch (eds.), *The Social Construction of Technological Systems. New Directions in the Sociology and History of Technology* (Cambridge, Mass.: MIT Press, 1987), pp. 51–82.

[4] T.P. Hughes, *Networks of Power. Electrification in Western Society 1880–1930* (Baltimore: Johns Hopkins University Press, 1983).

well as types of behaviour. They give rise to representations and patterns of thought, which relate, at various levels, to fundamentally different kinds of realities. For example, there are very general representations of efficiency, based on associated interpretations of nature and society, and there are more specific representations of the organization and the techniques of production. Mixing energy with social reflexions, the modern concept of work is a typical product of the first kind of representation.[5] There are also patterns that apply to the supervision of the workforce, while others again influence problems of design.

The main purpose of my work on technological thought is to suggest a close relationship between the collective mental frames I have described and the notions of coherence on which Gille's technological systems are founded. At the same time, the mental frames that I invoke necessarily relate foundations in technology to the forms of institutional and professional organization in which they become a material reality. They therefore play a role in the construction of systems of the kind that Hughes describes. It follows that their evolution is part of the pattern of major technological changes. In two recent books, I have tried to show how the transformation of the mental frames of engineers influenced the process that led France from the end of the classical age into the industrial era.[6] The main part of this article will deal with that specific example and with the lessons that can be drawn from it. But first, to clarify my approach, I wish to set out some general views on technology and technological thought.[7]

## TECHNOLOGY, NATURE, AND SOCIETY

Many historians, from David Landes to François Caron, define technology as a form of social production. It must not be forgotten, however, that technology deals with nature. More precisely, it provides an interpretation of nature in relation to the social division of labour. Technology is meant to adapt this division of labour to the principles of efficiency that can be deduced from observation of the physical world. These principles are partly the result of mental constructions and representations. There is no universal nature and there are no enduring principles of efficiency; rather there are historically determined

---

[5] See A. Rabinbach, *The Human Motor, Energy, Fatigue and the Origins of Modernity* (Berkeley and Los Angeles: University of California Press, 1992), as well as below.

[6] A. Picon, *L'Invention de l'ingénieur moderne. L'Ecole des Ponts et Chaussées 1747–1851* (Paris: Presses de l'Ecole nationale des Ponts et Chaussées, 1992). Some of the themes studied in this book are also dealt with in A. Picon, *French Architects and Engineers in the Age of the Enlightenment* (1988; English translation, Cambridge: Cambridge University Press, 1992).

[7] I have recently developed the theoretical bases of my approach to the history of technological thought and engineering in A. Picon, *Pour une histoire de la pensée technique. Rapport pour l'habilitation à diriger des recherches* (Paris, 1993).

representations of nature and of the principles of efficiency that derive from them.[8] Until the eighteenth century and the beginning of the first industrial revolution, for example, the principle of automatic working was synonymous with the transmission of movement: engineers such as the Renaissance designer Francesco di Giorgio conceived chiefly kinematic devices. During the nineteenth century, however, with the development of the steam engine, automatic working became synonymous with the production and transmission of energy.[9] It has changed again recently to correspond to the circulation of information. Through this kind of evolution, it seems that technology is constantly seeking to humanize nature by appropriating it to human needs and concerns. It attempts, at the same time, to naturalize society by adapting the supposedly natural principles of efficiency to the organization of production.

The history of technological thought must take account of the simultaneous representations of nature and society. As I would argue, the major transformations of technological thought correspond to changes in these representations. For instance, the evolution of engineering in eighteenth-century France was linked to a global change in the interpretation of the physical world and the construction of nature to which it led. The new vision of nature that emerged in the eighteenth century no longer focussed on architectonic regularities; instead it became increasingly centred on the natural dynamism of things and beings. The change was also linked to a new approach to society and to the role technology was supposed to play in human destiny. In the Enlightenment, the idea of collective progress was taking shape, and technological development and social progress seemed ever more closely related.

## TECHNOLOGICAL THOUGHT AND RATIONALITY

Technological thought is a complex system functioning at different levels. For each actor in the process of production, it comprises a mixture of know-how and interiorized rules of decision-making and action. Know-how plays a role in the worker's manual skill as well as in the engineer's design or in the manager's decision process. Technological thought also includes formalized knowledge, such as mechanics, physics, or chemistry for the engineer. And it embraces themes and representations that belong to an imaginary sphere; the common obsession of engineers with fluidity has clearly something to do with the imaginary. These various levels overlap. The ideal of automatic working appears, for example, at the intersection of the known and the imaginary.

This analysis is mainly concerned with collective actors, professional types, or professions. In other words, the history of technological thought that I should

---

[8] In this respect, I subscribe entirely to the approach developed by contemporary sociology of knowledge, notably (in the French context) by Michel Callon and Bruno Latour. See M. Callon and B. Latour (eds.), *La Science telle qu'elle se fait. Anthologie de la sociologie des sciences de langue anglaise* (Paris: La Découverte, 1991), pp. 29–35.

[9] J.-P. Séris, *Machine et communication* (Paris: Vrin, 1987).

like to promote is more concerned with shared know-how, knowledge, and representations than it is with the content of the individual mind. Applied to the question of invention, it means that the history of technological thought, as I conceive it, has more to do with the mental context that allows invention to take place, than with the actual process of invention. For example, if we take a single actor in a production process, let us say an engineer, the main problem is how to describe the complex system of that actor's technological thought, using appropriate interpretative structures or concepts. The historian must also fashion other structures or concepts in order to characterize the common points of reference that different actors share within large firms or institutions. In order to remain concise, I shall discuss only the first of these problems.

One can use the concept of rationality to clarify the specific organization of know-how, knowledge, and representations that characterizes the technological thought of the actors of production. The definition of rationality to which I refer here is very different indeed from the classic definition, especially as used in economics. In the first place, it does not separate the objectives from the means employed, in the way that Max Weber did when he distinguished between rationality depending on the objectives and rationality depending on the means.[10] In my definition of rationality, objectives and means are in constant interaction with each other in a process of technological reflexion and action. Secondly, my definition does not limit rationality to rational calculation. It characterizes the global attitude of an actor towards the world of production, rather than the specific techniques of decision he uses to cope with it. Rationality, therefore, is guided by values and representations. Some of these values and representations are ordinarily considered as irrational. In this sense, rationality is a general disposition linking all levels of technological thought and providing a frame for schemes of rationalization. Technical thought and rationality are linked to the devising of schemes or plans, and it is this planning dimension which gives them a dynamic quality.

Looking at the history of a number of forms of engineering rationality from the Renaissance to the present, we can distinguish successive ages. For example, there seems to have been an age of classical, i.e. geometrical or Vitruvian, rationality. At the turn of the eighteenth and nineteenth centuries, this age was supplanted by another that could be described as the age of analytical rationality. Focussing on the case of French engineers, I now turn to this transition from geometrical or Vitruvian rationality to analytical rationality.

## FROM GEOMETRICAL TO ANALYTICAL RATIONALITY

Since I have emphasized the importance of representations of nature, I begin by evoking the change in the perception of the physical world that took place in eighteenth-century France. In that period, there was a general trend towards a more dynamic perception of nature. Whereas the philosophers, scientists, and

---

[10] M. Weber, *Economie et société* (1921; French translation, Paris: Plon, 1971).

engineers of the classical period had considered nature as something organized according to the laws of order and proportion, i.e. as something essentially architectonic, eighteenth-century élites were increasingly impressed by the mobility of natural elements. In the seventeenth century, Bossuet wrote that God had created the world with order and proportion, but for Diderot and D'Holbach, change and movement were the main characteristics of nature.[11] Mobility was now seen as synonymous with a vital activity, while immobility became synonymous with decay and death. Efficiency was no longer linked to an ideal arrangement of means ruled by proportions; it was seen as the expression of natural dynamism. Eighteenth-century urbanism and territorial planning were clearly inspired by this conception. Nothing should prevent water and air from circulating freely in cities. As Parmentier put it in his *Dissertation sur la nature des eaux de la Seine* of 1787, "It is well known that the healthiest water would corrupt without the movement which maintains its purity".[12] The elimination of urban ditches, with their stagnant water, was a consequence of this kind of conviction. The destruction of houses built on bridges was another, since houses in such a position were supposed to prevent the renewal of the atmosphere.[13] Understanding natural dynamism was becoming a very general concern of philosophers and scientists as well as a practical issue for architects and engineers.

If natural efficiency was linked to mobility, social happiness and progress had to lie in a similar mobility, embracing the mobility of people, ideas, and merchandise. The fight against the prejudices that prevented free exchanges between people, and the promotion of trade and free enterprise, through the removal of customs barriers as well as the elimination of the corporate organization of labour, became the priorities of political and economic élites. The Physiocrats and their famous formula "laissez faire, laissez passer" illustrate that concern very well.

French engineers clearly subscribed to this ideal of natural and social fluidity, especially the Ponts et Chaussées engineers, whose mission was to build bridges and highways with a view to promoting trade. In relation to this concern, social evolution in at least two areas must be taken into account. First, engineers who

---

[11] See for instance J.-B. Bossuet, *Introduction à la philosophie, ou de la connaissance de Dieu et de soi-mesme* (Paris: R.-M. d'Espilly, 1722); D. Diderot, *Pensées sur l'interprétation de la nature* (1754); P.-H.-D. D'Holbach, *Système de la nature, ou des lois du monde physique et du monde moral* (1770; reissued Paris: E. Ledoux, 1821).

[12] A.-A. Parmentier, *Dissertation sur la nature des eaux de la Seine, avec quelques observations relatives aux propriétés physiques et économiques de l'eau en général* (Paris: Buisson, 1787), p. 21.

[13] On these major preoccupations of eighteenth-century urbanism, see, for example, B. Barret-Kriegel, B. Beguin, B. Fortier, D. Friedmann, and A. Monchablon, *La Politique de l'espace parisien à la fin de l'Ancien Régime* (Paris, 1975); J.-C. Perrot, *Genèse d'une ville moderne. Caen au XVIII^e siècle* (Paris and The Hague: Mouton, 1975); A. Guillerme, *Les Temps de l'eau. La cité, l'eau et les techniques* (Seyssel: Champ Vallon, 1983); B. Lepetit, *Les villes dans la France moderne (1740–1840)* (Paris: Albin Michel, 1988); Picon, *French Architects and Engineers*.

had previously been relatively marginal came to be recognized as important agents of economic and social progress. In France, this recognition took place under the patronage of the state, through the creation and development of specialized institutions, administrative corps, and professional schools. While the corps of fortification engineers dates from the end of the seventeenth century, the Ponts et Chaussées and the mining corps were founded during the eighteenth century.[14] In addition, a whole range of schools was created: the Ecole des Ponts et Chaussées in 1747, the Ecole du Génie at Mézières (a school for fortification engineers) in 1748, the Ecole des Mines in 1783, and the Ecole Polytechnique in 1794. Gaining rapidly in prestige, these schools were to play a decisive role in winning recognition for engineers and the rise in their status.[15]

As all this happened, the attitude of engineers towards society began to change. Engineers defined themselves less and less as artists serving a prince, on the model of the architect-engineers of the Renaissance and the classical age. They considered themselves rather as responsible for a more collective form of progress; and they saw it as their duty to defend new values, such as public utility and prosperity. "It is the engineer who is in charge of the designs meant to provide happiness"[16] was the comment of one Ponts et Chaussées engineer convinced of the extreme importance of transport infrastructures.

At the intersection of the ideal of fluidity and the evolution of the engineering profession, a gradual but radical change in the engineers' field of competence took place. The knowledge possesed by engineers and the way in which they conceived and designed their projects were destined to change. Engineers no longer defined themselves by reference to their mastery of purely geometrical knowledge, as designers or as "artist engineers" closely related to architects. For that purpose they created a new science, involving the use of calculus. As part of this change, they adopted new spatial and constructive patterns and, in turn, new methods of design.

Until the end of the eighteenth century, this change in the profile of engineering competence was impeded by a very traditional technological context, by a "classical technological system", as Bertrand Gille would say. It was also prevented by a system of knowledge that was still based on Vitruvian principles and

---

[14] On those different corps, see, for instance, A. Blanchard, *Les Ingénieurs du "Roy" de Louis XIV à Louis XVI. Etude du corps des fortifications* (Montpellier: Université Paul Valéry, 1979); J. Petot, *Histoire de l'administration des Ponts et Chaussées 1599–1815* (Paris: M. Rivière, 1958); A. Thépot, "Les ingénieurs du Corps des Mines au XIXᵉ siècle. Recherches sur la naissance et le développement d'une technocratie industrielle", doctoral thesis (Paris, 1991).

[15] R. Taton (ed.), *Enseignement et diffusion des sciences en France au XVIIIᵉ siècle* (Paris: Hermann, 1964); C.C. Gillispie, *Science and Polity in France at the End of the Old Regime* (Princeton, NJ: Princeton University Press, 1980); Picon, *L'Invention de l'ingénieur moderne*.

[16] P. Planier, French essay of 1779, manuscript in the library of the Ecole nationale des Ponts et Chaussées, Carton "Concours de style et concours littéraires 1778–1812".

the intensive use of geometric patterns. What the most advanced engineers tried to achieve, however, was an understanding and mastery of natural and human process ranging from floods to the organization of labour on construction sites or in factories. As they lacked the scientific and technical tools that would enable them to control these diverse realities, they used a provisional method, consisting of a systematic decomposition of things and phenomena, a decomposition that was intended to lead, in due course, to a more rational recomposition. It must be noted that, in accordance with eighteenth-century political philosophy, they also interpreted society as something that could be decomposed into individuals before being recomposed in terms of institutions and nations. Among the works that were inspired by this conception was a *Traité des richesses* by the Ponts et Chaussées engineer Achille-Nicolas Isnard. In this treatise, published in 1781, Isnard defined the "science of man" as a kind of mechanics based on the rational decomposition and recomposition of individual interests.[17] Territory as well could be decomposed and recomposed. The creation of the departments of France at the beginning of the Revolution was precisely the result of a decomposition of the old system of regions and its replacement in a process of rationalized recomposition.[18] Engineers, however, applied this method mainly to technical devices, as well as to the process of production, which they perceived as a combination of the individual activities (or "gestes") of workers — a combination that would yield more complex, integrated technical "operations". This attitude is well illustrated by their approach to engineering and architectural design in terms of basic functions and movements. Once identified, these functions and movements provided a general frame for the actual design of the equipment — a bridge or a building, for example. This approach is also detectable in their studies of human labour, notably in Charles-Augustin Coulomb's famous memoir on "The quantity of action men can develop by their daily work, according to the different ways they exert their strength".[19]

---

[17] On Isnard, see J.-C. Perrot, "Premiers aspects de l'équilibre dans la pensée économique française", *Annales. Economies sociétés civilisations* (1983), **5**: 1058–74.

[18] See M.-V. Ozouf-Marignier, *La Formation des départements. La représentation du territoire français à la fin du 18e siècle* (Paris: Editions de l'Ecole des Hautes Etudes en Sciences Sociales, 1989).

[19] C.-A. Coulomb, *Résultats de plusieurs expériences destinées à déterminer la quantité d'action que les hommes peuvent fournir par leur travail journalier, suivant les différentes manières dont ils emploient leurs forces* (Paris, 1799; reissued in C.-A. Coulomb, *Théorie des machines simples*, Paris: Bachelier, 1821, pp. 255–97). On Coulomb's memoir, see C. Stewart Gillmor, *Coulomb and the Evolution of Physics and Engineering in Eighteenth Century France* (Princeton, NJ: Princeton University Press, 1971); M. Valentin, "Charles-Augustin Coulomb (1736–1806)", *Sécurité et médecine du travail* (1974–5), **33**: 19–26; F. Vatin, *Le Travail. economie et physique 1780–1830* (Paris: Presses Universitaires de France, 1993).

This method was very similar to the one used in the descriptions of the arts and crafts in the *Encyclopédie*. The convergence is not surprising, since many encyclopaedists shared the same concern for rationalization. In his article on the knitting machine, Diderot laid bare the principles of a satisfactory description. It was based on "a kind of analysis which consists in distributing the machine in various parts ... before assembling those parts to rebuild the entire machine".[20] Manufacturing processes could be decomposed and recomposed in the same way. The engineer Jean-Rodolphe Perronet, who was later to become the first director of the Ecole des Ponts et Chaussées, gave a remarkable demonstration of that possibility when he studied a pin factory in 1739–40.[21]

This procedure also bears some analogy with the very general analytical method used by philosophers at the time. In his *Cours d'études* of 1775, Condillac wrote: "Analysis is the entire decomposition of an object and the arranging of its components so that generation becomes at the same time easy and understandable."[22] In this broad sense, we can speak of the emergence of a new kind of rationality which can be called analytical. It was linked to a novel way of studying efficiency in natural and social processes, founded on new relations between the parts and the whole, and between the local or the instantaneous and the global. Whether characterizing a natural phenomenon or an artefact as a machine, parts were supposed to combine in a dynamic way, instead of being composed according to rules of order and proportion analogous to the principles of architecture. What was actually at stake was the transition from a

---

[20] D. Diderot, "Bas", *Encyclopédie, ou dictionnaire raisonné des sciences, des arts et des métiers* (Paris, Briasson, 1751–1772), vol. 1, pp. 98–113, on p. 98. On this article, see J. Proust, "L'article 'Bas' de Diderot", in M. Duchet and M. Jalley (eds.), *Langue et langages de Leibniz à l'Encyclopédie* (Paris: 10/18, 1977), pp. 245–71. About the analytic method used by the encyclopedists, see also A. Picon, "Gestes ouvriers, opérations et processus techniques. La vision du travail des encyclopédistes", *Recherches sur Diderot et sur l'Encyclopédie* (1992), **13**: 131–47.

[21] J.-R. Perronet, "Explication de la façon dont on réduit le fil de laiton à différentes grosseurs dans la ville de Laigle", 1739, manuscript in the library of the Ecole nationale des Ponts et Chaussées, MS 2383; J.-R. Perronet, "Description de la façon dont on fait les épingles à Laigle en Normandie", 1740, manuscript in the library of the Ecole nationale des Ponts et Chaussées, MS 2385. Some of the drawings made by Perronet on this occasion were later used in the *Encyclopédie*.

[22] E. Bonnot de Condillac, "Cours d'études pour le prince de Parme", "V. De l'art de penser", in *Œuvres philosophiques de Condillac* (Paris: Presses Universitaires de France, 1947–1951), vol. 1, p. 769. On Condillac's definition of analysis and the later interpretations given to this key term in eighteenth-century philosophy, see J.-G. Granger, *La Mathématique sociale du marquis de Condorcet* (1956; reissued Paris: O. Jacob, 1989); E. Brian, "La foi du géomètre. Métier et vocation de savant pour Condorcet vers 1770", *Revue de synthèse* (1988), **1**: 39–68.

static understanding of structures to a more dynamic understanding of operations and functions. Nature and society were taken to be organized through operations and functions that were synonymous with mobility. At the same time, eighteenth-century scientists were also examining operations and functions. Lavoisier's chemistry was clearly analytic, in the same sense as Diderot's descriptions or Condillac's theory of ideas and language. In their own field, engineers tried to put this promising analytic orientation into practice, although the results to which it led to were initially quite modest.

Instead of entering into further detail about this turning point in the history of engineering rationalities, I should like to evoke some of its main consequences. The first consists in the emergence of a new engineering science based on calculus. From the first half of the nineteenth century, mathematical analysis progressively replaced many of the geometrical tools that engineers used during the classical period. As early as the 1820s, Claude Navier's or Jean-Victor Poncelet's lectures at the Ecole des Ponts et Chaussées and the Ecole du Génie et de l'Artillerie at Metz were based on calculus and on derivations and integrations that linked local or instantaneous laws to global phenomena.[23] Spectacular progress was made, by this approach, concerning the strength of materials and applied hydrodynamics. In his *Résumé des leçons données à l'Ecole des Ponts et Chaussées sur l'application de la mécanique* of 1826, for example, Navier made a notable study of beams on supports as well as of beams with ends built-in to the structure. With proper integration, the expression of the local bending moment enabled him to give the formula of the global deflexion curve.[24] The frame of mind created during the eighteenth century and the new type of relationship between the parts and the whole that it induced, laid the foundation for this massive introduction of calculus in the field of technology.

The new engineering science that was encapsulated in the work of Navier or Poncelet displayed a clear continuity with the pure sciences that use calculus also. Whereas most Vitruvian engineers subscribed to the idea that there was a gap between science and engineering that could only be filled with a lengthy immersion in technical reality, their successors saw an intimate relation between science and technology. During the nineteenth century, technological education, focussed on an apprenticeship in engineering science (as opposed to one founded on practice) developed in accordance with this principle.[25]

---

[23] A. Picon, *L'Invention de l'ingénieur moderne*, p. 469–505.

[24] Cl. Navier, *Résumé des leçons données à l'Ecole des Ponts et Chaussées sur l'application de la mécanique à l'établissement des constructions et des machines* (Paris: F. Didot, Carilian-Gœury, 1826).

[25] On the development of technological education in France and in Europe, see for instance J. H. Weiss, *The Making of Technological Man. The Social Origins of French Engineering Education* (Cambridge, Mass.: MIT Press, 1982); A. Picon, *L'Invention de l'ingénieur moderne*; R. Fox and A. Guagnini (eds.), *Education, Technology and Industrial Performance in Europe, 1850–1939* (Cambridge: Cambridge University Press, and Paris: Editions de la Maison des Sciences de l'Homme, 1993).

This new form of engineering science defined itself as a science of natural processes. At the same time, it established strong links with another new kind of knowledge concerning human processes: that of economic calculation. It is no coincidence that engineers such as Claude Navier or Jules Dupuit rank among the pioneers of this form of economic knowledge.[26] Through the links between engineering science and economic calculation, one can apprehend another important aspect of the age of the analytical form of rationality. The preoccupation of the engineers was no longer the mastering of space alone, as was the case during the classical period; it lay now in the simultaneous mastering of space and time, the two dimensions being ordered in accordance with economic principles. In a sense, classical engineers knew only about book-keeping. Their thinking was bounded by the present on one side and by eternity on the other. They either had the means to act immediately, or they had to wait for an undefined time. The minuteness of their tasks was counterbalanced by the ideal of eternal monuments of technology comparable with the great works of the Romans. The new vision that was emerging, in fact, was the medium-term perspective of modern economics.

Last but not least, a new approach to the processes of production also appeared. Production was now seen, as I have said, as a combination of elementary moves organized into a succession of technical operations. Introduced by engineers such as Coulomb, this new grid of interpretation enabled the engineer to rationalize production by exposing its components, workers' actions, and elementary operations, and submitting them to the principles of science. In connexion with this conviction, a kind of "proto-Taylorism" can be identified among French engineers in the second half of the eighteenth century.[27] This "proto-Taylorism" reached its climax with the "revolutionary productions" of powder, weapons, and guns during the Terror.[28] Although its practical results were a failure in the short term, it gave birth to the modern notion of work. Used to find a common measure between the various gestures and operations engineers deal with, above all between human effort and mechanical power, the notion received its mathematical expression during the first half of the nineteenth century, when economists stressed its fundamental importance in the task of understanding social and economic processes.[29]

From the coupling of science and technology to the stress put on the notion of work, the emergence of analytical rationalities corresponded to the construction of a new collective mental frame. Some features of this new mental frame, such

---

[26] F. Etner, *Histoire du calcul économique en France* (Paris: Economica, 1987).

[27] Picon, *L'Invention de l'ingénieur moderne*; A. Picon, "Gestes ouvriers, opérations et processus techniques".

[28] C. Richard, *Le Comité de Salut Public et les fabrications de guerre sous la Terreur* (Paris: F. Rieder et Cie, 1922); P. Bret, "La pratique révolutionnaire du progrès technique. De l'institution de la recherche militaire en France (1775–1825)", doctoral thesis (Paris, 1994).

[29] I. Grattan-Guinness, "Work for the workers: advances in engineering mechanics and instruction in France, 1800–1830", *Annals of Science* (1984), **41**: 133; Vatin, *Le Travail*.

as the engineers' ideal of fluidity, are still influential in the industrial world today. At the same time, the reflexions on the transition from Vitruvian to analytical rationalities that I have been summarizing are inseparable from a broader programme of research. This programme is linked to the changes we are experiencing in our own day in the field of technology and industrial organization. Are we on the brink of another transition, between the analytical ideals we have inherited from the industrial revolution and a quite new conception of efficiency? When we consider the recent changes in our interpretation of nature or take account of some of the changes induced by the use of computers, we see that such a transition may be about to occur. Until now, efficiency was synonymous with the rational regulation and management of the dynamism created by mass production and consumption. But now, the complexities of our world and its uncertainty seem to call for a more systemic efficiency based on the identification of levels of technological and managerial relevance. How are we to articulate those different levels? In some advanced fields, engineers and managers are looking for new design and decision procedures taking into account the effects of complexity and uncertainty. Computer-assisted in most cases, these procedures try to order logically the possible events that can affect production and marketing. Whereas the analysis of operations and functions was clearly the basis of former dynamic regulations, the use of logical structures may be the mark of a new approach to technological efficiency that is emerging in our own day.[30]

## TECHNOLOGICAL THOUGHT AND CULTURE

In conclusion, I offer two further comments on the history of technological thought as I have tried to present it. The first concerns the definition of the engineer. In a recent article, the American sociologist Peter Whalley rejects what he calls "essentialist definitions" because of the extreme diversity of engineers' areas of competence and employment.[31] To cope with this diversity, he suggests that we should define engineers by their status — as privileged employees trusted by the owners of capital and by the managerial interest. The attraction of such an approach is undeniable. At the same time, I think it is necessary to qualify the approach by taking into account the history of engineers' rationalities. After all, engineers can also be regarded as being in constant pursuit of a systematic form of knowledge whose aim is a union of the organization of production with efficiency. The historical forms which this pursuit takes may perhaps throw some light on the possible definition of the profession.

A second remark concerns the relationship between the history of the rationalities of engineers and social history. It must be clear that this relationship

---

[30] Picon, *Pour une histoire de la pensée technique*.

[31] P. Whalley, "Negotiating the boundaries of engineering: professional managers and manual work", *Research in the Sociology of Organizations* (1991), **8**: 191–215.

is mainly achieved through the mediation of culture. The history of rationalities provides a way of connecting technological change with cultural evolution. Such a connexion is in no way determinist; in other words, the transformation of technological thought is not always a necessary condition for technical innovation. The steam engine, for instance, did not develop because a new frame of mind, leading to an analytical engineering science, emerged during the eighteenth century; rather, engineering science was stimulated by the development of the steam engine. However, it can easily be demonstrated that, without a new analytical frame and the scientific results that this frame yielded, the development of machinery would eventually have stagnated. As this example shows, instead of a direct causal link between technological thought and innovation, one is often faced with indirect or deferred causality. Lewis Mumford pointed out the interest of this kind of causality when he saw the distant roots of the possibility of mechanization in the precise measurement of time introduced in medieval monastic life.[32] On a smaller scale, the history of engineering rationality contributes to explanations of the same type. It helps us to understand the meaning of technological change before trying to assign definite causes to it.

---

[32] L. Mumford, *Technique et civilisation* (1934; French translation, Paris: Seuil, 1950), pp. 22–6.

# Bodies, Fields, and Factories: Technologies and Understandings in the Age of Revolutions

## John V. Pickstone

Our history tends to be compartmentalized: industrial vs pre-industrial; sciences vs technology; the physical vs the bio-medical. This paper is deliberately synthetic. It sets out to problematize these divisions and offers some new perspectives on how the divisions might be bridged.

I concentrate here on the eighteenth and nineteenth centuries and I focus on understandings, both of nature and of the works of man. I take together the "physical" and the "biological". I try to draw on the history of biology and medicine, as well as on the history of physical sciences and technology. I also use, and hope to enlarge, links between historians and those anthropologists who are interested in material culture.

I present two types of understanding — two ways of understanding, which are also types of scientific work. I refer to them as "savant" and "analytical".[1] The former I see as based on natural history, on wide-ranging taxonomies using superficial characteristics, and as especially characteristic of the eighteenth

---

[1] In several recent papers I have begun to explore and use a historical typology of ways of knowing; I am now developing these ideas in book form. See J.V. Pickstone, "Ways of knowing: towards a historical sociology of science, technology and medicine", *The British Journal for the History of Science* (1993), **26**: 433–58; "Museological science? The place of the analytical/comparative in nineteenth-century science, technology and medicine", *History of Science* (1994), **32**: 111–38; and "Objects and objectives: notes on the material cultures of medicine", in Ghislaine Lawrence (ed.), *Technologies of Modern Medicine* (London: Science Museum, 1994), pp. 13–24.

century, but as continuing thereafter into our present.[2] Analytical understandings also existed in the eighteenth century[3] but were characteristic of the early nineteenth. They involved the real or cognitive decomposition of natural or manmade objects into "elements", which were specific to, and hence constitutive of, new sciences. Analytical chemistry was the usual model.

In this paper, I concentrate on artefacts and argue that they were often understood in the way I characterize as savant. I then show how such artefacts came (also) to be understood in analytical ways. I am especially interested in crafts, in agriculture and husbandry, and in mining and medicine — topics that tend to fall between the history of technology and the history of science. I try to present these technical objects as part of cultural history, without resorting to *zeitgeists* or *epistemes*, floating in cultural or linguistic spaces.[4] Rather I link these understandings to specific innovations in élite science, and I seek to ground them in social and economic history, specifically in the transformations associated with the French Revolution and the Industrial Revolution.

My major historiographical resource is Michel Foucault, especially his earlier books, which have not been used much by serious scholars, in spite of (or perhaps because of) the recent vogue for his later works on sex and discipline. I want to extend Foucault's account of eighteenth-century understandings of plants, grammar, and wealth, and then to show how we might understand the emergence of characteristically nineteenth-century knowledges in terms of social and economic history, without resorting to the kind of intellectualist catastrophism that limited Foucault's own treatment of biological sciences (but not his treatment of medicine).[5]

---

[2] The best guide to natural-historical understandings in the eighteenth century remains Michel Foucault, *The Order of Things. An Archaeology of the Human Sciences* (London: Tavistock, 1970), and *The Birth of the Clinic. An Archaeology of Medical Perception*, trans. A.M. Sheridan (London: Tavistock, 1973).

[3] In planetary astronomy and rational mechanics, bodies of data were analysed by "mathematicians" — professional exponents of esoteric knowledges considerably removed from usual savant forms.

[4] In *The Order of Things*, Foucault presented eighteenth-century botany, etc. as part of a classical *episteme*; this has sometimes been interpreted as a form then taken by all organized knowledge, one that was prior to all particular understandings. The classical *episteme* was supposedly displaced at the end of the century, for reasons that Foucault did not attempt to explain. For an excellent explication of early Foucault, see Gary Gutting, *Michel Foucault's Archaeology of Scientific Reason* (Cambridge: Cambridge Univesity Press, 1989). I want to link ways of knowing to social structures and to argue for plurality at any given time. Such a method links knowledges to social history, as Foucault did for medicine (in *The Birth of the Clinic*) but not the sciences discussed in *The Order of Things* (note 2).

[5] See notes 2 and 4.

## SAVANTS, CONNOISSEURS AND ARTEFACTS

"Natural history" is easily dismissed by historians of science, as are agriculture and crafts by historians of technology. But economic history would warn us against that second rejection, as a social history of Enlightenment savants would warn us against the first. This is not just a matter of statistics, of what kinds of knowledge practices or economic practices were most important in the sense of being common; there is also an intellectual question. We need to focus on the cognitive significance of "natural-historical" understandings, whether of divine or human creations. We need to see how artefacts appeared in Enlightenment culture.

I want to suggest that artefacts, like plants, were usually understood in terms of types. Species or types had essential qualities and accidental qualities. Wild plants and animals could be understood in this way, and so could domestic varieties, for these too were often seen as natural kinds; traditional breeds of sheep, like wild plants or animals, were usually characteristic of particular localities. The argument can be extended to plant and animal products. Wines and cheeses we still tend to judge as more or less true to type; the varieties of wine and cheese are quasi-species, and more or less characteristic of particular regions. But the same frame can be and was applied to many crafts. Spades for cutting peat varied in form between the different dales of Yorkshire. Vernacular housing was seen, and often still is seen as a quasi-natural product which somehow grows out of a particular locality.

And as William Reddy reminded us in a splendid article on French descriptions of French textiles, eighteenth-century accounts were usually about specific local products, such as the *bleu de Nîmes*.[6] The elaborate national system of local guild production both perpetuated and documented local types, which merchants and inspectors recognized and described, characteristic by characteristic. Dictionaries of manufactures, it seems, were not so different from listings of animals and plants.

We can generalize this insight: craft production depended on experience and tacit knowledge, on traditions that were reproduced locally. The growth of supra-regional markets and the growth of "exploration" and "commentary" (especially through print) brought both local crafts and local "natures" into the comparative frameworks we are discussing here as natural history. The savant's knowledge was cosmopolitan, and so, often, was that of the connoisseur, who could recognize types of wine or types of people, and was also expert on their characteristic defects or diseases.

---

[6] William Reddy, "The structure of a cultural crisis: thinking about cloth in France before and after the Revolution", in A.Appadurai (ed.), *The Social Life of Things. Commodities in Cultural Perspective* (Cambridge: Cambridge University Press, 1986).

Historians of crafts and domestic arts may here like to draw on the history of medicine. For medicine, then, was the art of recognizing human types and human defects. Savant commentators on crafts or manufactures were probably doing much the same. As Charles Gillispie noted decades ago, the *Encyclopédie,* in its accounts of arts and crafts, was not a work of applied science, but rather a kind of natural history of industry.[7]

Our clue to understanding such knowledge/practices may well be the notion of patronage, or so the historiography of medicine would suggest.[8] But that is another story. Instead, from that preliminary characterization of savant/ connoisseur practices, I would move to a corresponding characterization of the knowledge practices that were mostly new in the nineteenth century. I want to argue that these involved novel forms of understanding, no longer about types but about analysis into elements.

## ANALYSIS AND ARTEFACTS

The model science here was analytical chemistry, and that was also the key link between new knowledge and new economic practices. But the new form of knowledge was very extensive: analytical chemistry was paralleled by new kinds of zoology, botany, and pathology, and several new kinds of anatomy, all of which had their own elements (for example, the tissues of general anatomy). In the new geology, rock formations were analysed into strata, and minerals into crystal forms. We can also argue for various kinds of "analytical physics" — new sciences of work, heat, light, electricity, etc. All these new sciences had (and have) domain-specific elements which served to explain both form and process. Indeed, natural forms (or species) now began to appear as products of processes, governed by laws; as points in a matrix of possibilities, no longer themselves the primary objects of knowledge.[9]

It seems to me important to recognize the cognitive similarities between these many new fields, even if I do not have room here to develop this aspect of the argument. Instead, I want to discuss how and why these knowledges were

---

[7] C.C. Gillispie, "The natural history of industry", in A.E. Musson (ed.), *Science, Technology and Economic Growth in the Eighteenth Century* (London: Methuen, 1972).

[8] N.D. Jewson, "Medical knowledge and the patronage system in eighteenth-century England", *Sociology* (1974), **8**: 369–85.

[9] See my "Museological science?", cited in note 1. The new physical sciences, we may suggest, construed their objects not as combinations of different elements but as systems through which particular elemental "fluids" flowed in ways which could be described mathematically. Thus Lazare Carnot followed mechanical impulse; Sadi Carnot, heat; Fresnel, light; Ampère, electricity. I am grateful to Jerry Ravetz for discussion on this point.

created. More specifically, I want to ask how these analytical understandings were related to various kinds of technical practices. Again I wish to range across a wide spectrum of practices and knowledges, ignoring some of the usual distinctions between science and technology. At one extreme, we are concerned with the analysis inseparable from certain re-orderings of capitalist production; at the other, with analyses produced chiefly for pedagogy, especially in new institutions of higher professional education. But these extremes may be better connected than most historians have understood.

At the first extreme, we could take, for example, English animal breeding. We have suggested that traditional breeds were usually seen as quasi-species, as in pedigree breeding they still are. But they were also subject to that capitalist reconstruction which the late E.P. Thompson described so well in his essay on the "Peculiarities of the English".[10] For Robert Bakewell, farming in eighteenth-century Leicestershire, a sheep could be a machine for turning grass into money, a compound of marketable traits, defined by costs and benefits. Types here have become secondary; and it is not accidental that a century later, with Darwin, the common view of wild species followed this same shift. The mediation between Bakewell and Darwin was the writings of commentators and improvers — more or less philosophical men and women who helped to disseminate the practices of non-writing breeders, making them part of "the new literatures" of "statistics" and political economy.

Perhaps we can see a similar shift in understandings of textiles. The article by William Reddy, to which I have already referred, shows how the decay of the guild system and the expansion of rural textile manufacture in the later eighteenth century rendered obsolete the dictionaries focussed on products. Eventually a new generation of texts was published, in which local "types" of cloth were unimportant; instead, the focus was on production. Whereas methods had once been traditional, local (even secret), and assimilated into the product-type, now production methods were publicized, and products were defined by market demands.

The key shift here, as in animal breeding, was not "industrialization", but rather a kind of analytical rationalization which was logically prior to industrialization. The crucial factor was a change in social relations of production — and a change in the relations of texts and processes. The new authors were not inspectors of guild regulations; they were exponents of political economy or analysts of machines.

In neither of these cases were the new forms of production matters of formal education, although the analytical knowledges were later institutionalized in agricultural and technical colleges. Indeed, we have here a situation which

---

[10] E.P. Thompson, "The peculiarities of the English", in his *The Poverty of Theory and Other Essays* (London: Merlin Press, 1978); Isabel A. Phillips, "Concepts and methods in animal breeding, 1770–1870" (Ph.D. thesis, Manchester: UMIST, 1989).

seems paradigmatic for much history of technology, especially in Britain: shifts in patterns of production, carried out by inventors or entrepreneurs, and then described by commentators in whose writings the new understandings are clarified and formalized. In some ways, that process is so familiar that we may fail to register the shifts of meaning. Here I wish to stress the change in understanding, from a type-type, as it were, to an analytical-type (and the associated change in the relation of "doers" and "describers"). I wish to explore the commonalities and connexion between these reconceptualizations of artefacts and the contemporary emergence of new sciences in and around new institutions of professional education.

## ANALYSIS IN MEDICINE AND ENGINEERING

Of the professional technologies in this period, medicine is probably the best-explored, and we can use this historical literature to consider both the production of treatments and the production of new systems of knowledge. Whereas the treatments available probably changed but little c.1800 (or indeed for decades thereafter), modes of medical teaching changed substantially, and important analytical understandings emerged, partly as a result of the French Revolution. In some Parisian hospitals, where charity-care had been "nationalized" and handed over to doctors, patients became objects of observation and experiment, specimens which doctors controlled; they were the material of a new kind of medical education. Patient biography and individual variations became less important, and diseases were no longer seen as aspects of the natural history of particular localities. There was no longer cause to worry that hospitals confused matters by co-mingling patients and diseases of many types and places. Instead, as in the case of animal breeding or textiles, local variety moved out of focus, and the bodies of paupers became mere examples of those universal, biological laws of production that governed normal and pathological structures. The analysis was conducted in terms of tissues; it depended on "large collections" of living and dead specimens and the right of doctors to take (dead) specimens apart. Hospitals were now museums of diseases, and they included museums of pathological anatomy. There were strong parallels and direct connections with the botanical gardens and the herbaria, the menagerie and anatomy galleries of the Museum of Natural History in Paris, a key site for the new sciences of organization, including chemistry and geology as well as botany, zoology, and morphology.

This new kind of medical understanding became widespread in the nineteenth century, and it persists still in much "clinical medicine". It provided improved diagnosis and the possibility of finer prognoses. It was the parent of most later techniques of analytical diagnosis, the intellectual basis of various kinds of medical and surgical specialization.[11] The relation of such analyses to the private practice of medicine is complex and interesting. To much general practice it was marginal, but, in France at least, the recognition among students and teachers that could be gained by excellence in pathological anatomy could be translated into state appointments and thence into successful private practice.[12] Medicine has a rather odd kind of market.

The analysis of corpses by élite French doctors may seem far distant from the mute capitalist rationalizations that were mentioned earlier. Yet there were links. We know that the new forms of analysis in medicine were based on surgical approaches — knowledges that were previously of low status and matters for craftsmen rather than savants.[13] New professional schools, often on politically radical principles, helped to elevate, extend, and formalize these practical knowledges within the new frames of analysis.

This is my cue for a move to engineering. Machines have seemed miles from medicine, but I wish to stress the similarities. As analytical medicine drew on formalized surgical knowledge, so, I would argue, the new analytical forms of physical sciences, drew on the previously low-status knowledges of engineers. I rely here on that recent, and not so recent, historiography of physics which has paid attention to social and economic contexts. For example, the history of "energy" in British physics, as recounted by Donald Cardwell and Arnold Pacey, suggests such a pattern.[14] So, it seems, does the French tradition which included Lazare Carnot on mechanical transmission, and Sadi Carnot and Joseph Fourier on heat. This theme again emerges for France in the recent study by Crosbie Smith and Norton Wise, although their social history is unclear until they focus on Glasgow and Lord Kelvin.[15] My general model would direct attention to the recurrent French attempts, e.g. in the early Ecole Polytechnique, to find

---

[11] See my "The biographical and the analytical: towards a historical model of science and practice in modern medicine", in I. Löwy *et al.* (eds.), *Medicine and Change. Historical and Sociological Studies of Medical Innovation* (Paris: Editions INSERM, John Libbey, 1993).

[12] I am grateful to Patrice Pinell, who pointed out that the majority of the prestigious doctors in early-nineteenth-century France made most of their money in private practice, not from state salaries (as most historical studies seem to imply).

[13] T. Gelfand, *Professionalising Modern Medicine. Paris Surgeons and Medical Sciences and Institutions in the Eighteenth Century* (Westport: Greenwood Press, 1980); O. Temkin, "The role of surgery in the rise of modern medical thought", *Bulletin of the History of Medicine* (1951), **25**: 248–59.

[14] D.S.L. Cardwell, *From Watt to Clausius. The Rise of Thermodynamics in the Early Industrial Age* (London: Heinemann, 1971), and *James Joule. A Biography* (Manchester: Manchester University Press, 1989).

[15] C. Smith and M.N. Wise, *Energy and Empire. A Biographical Study of Lord Kelvin* (Cambridge: Cambridge University Press, 1989).

analytical principles of engineering and industrial practice, which could be taught in formal courses.[16] Historical studies by Robert Fox, Terry Shinn, and Ivor Grattan-Guinness are suggestive here.[17] It is in French professional education, whether for medicine or for engineering, that the rationalizations of practices and the rationalizations of philosophers were most intimately intertwined, precisely in such nationalized sectors as military engineering, state medicine, and state factories.

But let me pursue the parallels for another aspect of analytical engineering: the conceptual decomposition of complex machines into their elements — elementary machines. The classic text is by Franz Reuleaux, a mid-nineteenth-century German who drew explicit analogies between his studies of machines and the comparative anatomy of animals.[18] But the founding text on the analysis of machines was published by Lanz and Bétancourt in Revolutionary France.[19] The context was the new Ecole Polytechnique and the Ecole des Ponts et Chaussées. The parallel with medicine is very close. The Ecole Polytechnique, like the Ecole de Santé, was initially a radical scheme to teach the principles of practice, situated somewhere between "natural philosophy" and mere empiricism. As in medicine, collections played a major role; the schools had museums used in teaching. Indeed Agustín de Bétancourt was a Canarian sent by the Spanish Government to survey machines through Europe and to build accurate scale models, intended for a new Spanish school of engineering — a rather nice conjuncture of the museological and the analytical.[20] Here, very directly, the products of technology became the objects of a new form of analytical knowledge produced and disseminated as part of a new form of technological reproduction.

---

[16] Antoine Picon, *L'Invention de l'ingénieur moderne. L'Ecole des Ponts et Chaussées 1747–1851* (Paris: Presses de L'Ecole nationale des Ponts et Chaussées, 1992).

[17] R. Fox, "Laplacian physics", in R.C. Olby *et al.* (eds.), *Companion to the History of Modern Science* (London: Routledge, 1990), pp. 264–77; I. Grattan-Guinness, "Work for the workers: advances in engineering mechanics and instruction in France, 1800–1830", *Annals of Science* (1984), **41**: 1–33 and "The ingénieur savant, 1800–1830: a neglected figure in the history of French mathematics and science", *Science in Context* (1993), **6**: 405–433; T. Shinn, "Science, Tocqueville, and the state: the organisation of knowledge in modern France", *Social Research* (1992), **59**: 533–66; also see the essays by L.J. Daston on probability, by J.V. Grabiner on analysis, and by I. Grattan-Guinness on French mathematical physics in H.N. Jahnke and M. Otte (eds.), *Epistemological and Social Problems of the Sciences in the Early Nineteenth Century* (Dordrecht: D. Reidel, 1981).

[18] F. Reuleaux, *Kinematics of Machinery,* trans. A.B.W. Kennedy (London: Macmillan, 1876).

[19] J.M. Lanz and A. de Bétancourt, *Essai sur la composition des machines* (Paris, 1808).

[20] J.F. Perez and I.G. Tascon (eds.), *Description of the Royal Museum Machines* (Madrid: Ediciones Doce Calles, 1991).

In such ways, for engineering as for medicine, historians might avoid such crippling terms as "applied science". But can we understand in this way the scientific knowledges and practices that were philosophically and economically central at the opening of the nineteenth century — those of analytical chemistry? I think we can. Here I stick my neck out again, in the hope of dialogue with historians more familiar with the details.

## ANALYSIS AND THE ELEMENTS OF CHEMISTRY

Again we must move our focus from theory to practices and especially to practices of sufficient economic importance to support men who would call themselves chemists. The key practice, of course, was analysis, especially, it seems, the analysis of minerals and of medical materials. From the standard history of analytical chemistry you will discover that the key topic for this period was the analysis of mineral water, which seems odd, until you realize its economic importance and its position at the intersection of mineralogical and medical chemistry.[21]

The key country was Sweden. The key industry was Swedish mining. The key figure was Berzelius, who spent most of his career professing chemistry at the new School of Surgery in Stockholm. In a recent collection of essays on Berzelius, Evan Melhado has shown brilliantly how Swedish mineralogy moved from a kind of natural history (an excellent example of what I call savant knowledge) to become a matter of chemical (or crystallographical) analysis, where the properties of compounds were seen as emergent in a way characteristic of all later chemistry.[22] The creation of a systematic analytical chemistry was also a sophisticated re-ordering of minerals (and of medical materials), a new kind of understanding of economic practices. I should like to know much more about Berzelius, but he was undoubtedly central not just to chemical theory but to the rapidly expanding practice of analysis, for the first three decades of the nineteenth century. He fits the model rather well: the institutionalization of the new chemistry here appears as intimately linked to the practices of mining and medicine, to the associated museological collections, and to new forms of professional education. Minerals and organic materials were no longer to be

---

[21] F. Szabadvary, *History of Analytical Chemistry*, trans. G. Svehla (Oxford: Pergamon Press, 1966); C. Hamlin, *A Science of Impurity. Water Analysis in Nineteenth Century Britain* (Bristol: Adam Hilger, 1990).

[22] Evan M. Melhado, "Novelty and tradition in the chemistry of Berzelius (1803–1819)", in Evan M. Melhado and T. Frängsmyr (eds.), *Enlightenment Science in the Romantic Era. The Chemistry of Berzelius and its Cultural Setting* (Cambridge: Cambridge University Press, 1992), pp. 132–70.

classified by surface properties, as in savant natural history, they were to be taken apart, conceptually and physically, not least by that new instrument of chemical dissection, the galvanic battery. In major debates over the future of higher education in Sweden, Berzelius argued against the German models used to legitimate small-town academic communities; instead, he argued for Parisian models, where the new sciences were intimately related to professional education and technological practices.

The conjunctures become even more obvious with the key figure of the 1830s and 1840s — Justus Liebig, who turned a programme for the education of pharmacists into a world centre for an analytical chemistry that was at once medical, agricultural, and industrial.[23] Again we see the centrality of professional education and of collections. And here we return to economic history and to liberal polities not noted for state sponsorship of professional education: it is well established that the disciples of Liebig provided the analytical expertise which by the mid-nineteenth century was providing returns for British industry and American agriculture.[24]

In chemistry, as in other disciplines, analytical knowledges did not, in general, create new technologies — in fields, in factories, or in hospitals. Analysis was applied to existing processes and objects. The conclusions were often deductive — as in Liebig's agricultural chemistry, where, from analyses of soils and plants, you might deduce the best kind of fertilizer.

If the intellectual and "public" appeal of analysis often lay in display (e.g. in museums), its practical appeal lay in the regulation of technologies — the analysis of inputs and outputs, whether in factories or in patients. This is not "applied science" in any of the usual senses, but it was enormously important. As Nathan Rosenberg has often pointed out, and as Andrew Carnegie had discovered for himself, one could make an awful lot of money from the better control of inputs and outputs.[25]

# CONCLUSIONS

I have stressed analysis into elements as the cognitive aspect of many of the knowledge-practices and practical-knowledges developed around 1800 (although I also wish to underline the continuance of savant, natural-historical

---

[23] J.B. Morrell, "The chemist breeders: the research schools of Liebig and Thomas Thomson", *Ambix* (1972), **19**: 1–46; R.S. Turner, "Justus Liebig versus Prussian chemistry: reflections on early institute-building in Germany", *Historical Studies in the Physical Sciences* (1982), **13**: 129–62.

[24] M. Rossiter, *The Emergence of Agricultural Science. Justus Liebig and the Americas 1848–1880* (New Haven: Yale University Press, 1975).

[25] D.C. Mowery and N. Rosenberg, *Technology and the Pursuit of Economic Growth* (Cambridge: Cambridge University Press, 1989).

knowledges). I have related these new kinds of analysis to capitalist rationalization, picked up by commentators, and to professional educators, consciously creating new sciences. I have stressed the intermediate grounds, the analyses created in new schools which formalized economic practices such as surgery or engineering.

And if we are looking for new historiographical directions, then perhaps we can generalize some of the themes of this paper. We need to expand our understanding of technology as culture, but to avoid the inversions of cultural idealism. We need to expand our understanding of science as work, but to focus on "production" and the economy as well as on the work of rhetoric. Through the study of knowledge-practices and practical-knowledges we may learn to integrate the cognitive and the social aspects of history. We shall do so all the better for recognizing the synchronic and diachronic variety of sciences and technologies, and by talking about factories and fields and hospitals as well as about laboratories.

Lastly, as we continue to explore the specific fields, we might also ask ourselves about the big pictures, if only because we can all gain from historiographical cross-fertilization, risky though it may be.

# Evolution and Technological Change: A New Metaphor for Economic History?*

*Joel Mokyr*

## EVOLUTIONARY PARADIGMS AND DARWINIAN DYNAMICS

The recent attractiveness of the Darwinian mode of thinking to social scientists and historians is neither unexpected nor unusual. In discussions of institutions, technology, and social change, the fundamental intellectual power of Darwinian models has become increasingly influential in recent decades. Darwin's ideas have been extended from biology to such areas as information system dynamics and economic history.[1] Technological change lies at the intersection of

---

This paper is a much condensed version of a longer one, available from the author upon request. The financial assistance of National Science Foundation Grant SES 9122384 is acknowledged.

[1] See, for example, Philippe Van Parijs, *Evolutionary Explanation in the Social Sciences. An Emerging Paradigm* (Totowa, NJ: Rowman and Littlefield, 1981) and E.L. Jones, "Technology, the human niche, and Darwinian explanation," in E.L. Jones and V. Reynolds, (eds.), *Survival and Religion: Biological Evolution and Cultural Change* (Chichester: Wiley, forthcoming, 1995). For an application to the history of science, see David L. Hull, *Science as a Process* (Chicago: University of Chicago Press, 1988). Some well-known economists have also shown a growing interest in evolutionary issues: see, for example, Amartya Sen, "On the Darwinian view of progress", *Population and Development Review* (March 1993), **19**: 123–37; Martin Weitzman, "On diversity", *Quarterly Journal of Economics* (May 1992), **107**: 363–406; and Geoffrey Hodgson, *Economics and Evolution. Bringing Back Life Into Economics* (Cambridge: Polity Press, 1993).

information theory and economic history, and it is natural, therefore, to turn to Darwinism and to draw an explicit isomorphism between technological history and natural history. While instructive, however, analogies can be misleading. The evolutionary method in technological change is not so much an *analogy* with biology as another application of a Darwinian logic that transcends the world of living beings.[2]

Darwinian dynamics are based on two basic elements: random variation and selective retention. The "system" that Darwin was thinking of consisted of living beings belonging to larger groups called "species". The historical process of change consists of two separate processes: changes of the species themselves (mutation, speciation, and extinction), and changes in the distribution of different species among the population (selection). Darwinian selection is more than "survival of the fittest": it is the systematic retention of certain individuals and species according to predetermined criteria.

The mechanism is one of interest to historians of knowledge, whether technological, cultural, or scientific, because Darwinian mechanisms have been proposed to explain the change in epistemological systems. The main idea proposed by Campbell and his followers in the history of knowledge is that blind variation and selection mechanisms can be applied to abstract concepts as well as to living beings. Economic historians, more than most, should take note. "New ideas", however we define them, can be said to arrive in a highly stochastic manner, much like mutations. They result from a mysterious combination of social context, individual genius, and pure "luck". They are then subjected to merciless filtering processes that weed out the vast majority of them. The probability of success is a function of the selection criteria, but for a variety of reasons, sometimes seemingly favourable mutations are not selected.

The applicability of this intuitively powerful idea to the history of technology is not new. Many writers in various subspecialties have been struck by the common elements of technology and evolution. In the past few years, a number of economic and historians of technology, myself included, have tried to lay

---

[2] Darwin himself recognized the similarity of his evolutionary system and historical development of languages. Change occurs through blind variations, most of which will be rejected, but in some circumstances a small percentage will be retained, increasing the overall performance effectiveness of the system given its environment. Like species, some languages become extinct and others branch off existing structures, tantamount to speciation. Charles Darwin, *The Descent of Man and Selection in Relation to Sex* (New York: Modern Library edition, 1871), p. 466.

down the basic issues.[3] As I have argued elsewhere, evolutionary metaphors in social science should only be transplanted from their biological origins with great care.[4] The notions of blind variation in existing configurations and of selective retention mean somewhat different things in technological and biological contexts. For one thing, new technological information does not appear as random as it does in Weismannian genetics: it has a direction given to it by purposeful individuals intentionally trying to solve certain problems, it is subject to what is known as orthogenesis or "directed variation".[5] We may define a measure like a correlation coefficient that measures the dependence of mutations or inventions on the environment. Depending on this parameter, we are between the Darwin–Weismann world of pure independence and a "necessity is the

---

[3] The analogy has been suggested, among others, by Carlo M. Cipolla, "The diffusion of innovation in early modern Europe", *Comparative Studies in Society and History* (1972), **14**: 46–52; Brooke Hindle, *Emulation and Invention* (New York: New York University Press, 1981), p. 128; and Jane Jacobs, *Cities and the Wealth of Nations* (New York: Random House, 1984), pp. 223–5. For recent accounts by technological historians, see George Basalla, *The Evolution of Technology* (Cambridge: Cambridge University Press, 1988); Walter Vincenti, *What Engineers Know and How They Know It. Analytical Studies from Aeronautical History* (Baltimore, Md: Johns Hopkins University Press, 1990); Walter Vincenti, "Variation-selection in the innovation of the retractable airplane landing gear: the Northrop 'anomaly'" (unpublished manuscript, Stanford University, 1992); and Eric L. Jones, "The economic history of technology in Darwinian mood" (La Trobe University, School of Economics and Commerce, discussion paper 22.90, 1990). An economic approach with emphasis on systems theory is Norman Clark and Calestous Juma, *Long-run Economics. An Evolutionary Approach to Economic Growth* (London: Pinter Publishers, 1987).

[4] See Joel Mokyr, "Evolutionary biology, technical change, and economic history", *Bulletin of Economic Research* (April 1991), **43**: 127–49.

[5] The Weismann orthodoxy of complete independence between the forces that govern the direction of mutation and those that govern selection, and the orthogonality of mutation to environment still appears to come under challenge from time to time. Robert Wesson has argued that "the impossibility of input from the environment to the genome has never been proved. In view of the inventiveness of life it would be surprising if a law of nature absolutely prohibited it"; see Robert Wesson, *Beyond Natural Selection* (Cambridge, Mass.: MIT Press, 1991), p. 227. This is, of course, a far cry from anything remotely resembling proof. Recent work in molecular genetics, however, suggests that this kind of mechanism is not impossible. If it indeed turned out that the direction of mutation could be influenced by the environment so that "needed" mutations are more probable than others, another large difference between cultural and biological evolution would be removed. Recent work in molecular biology, while still in an embryonic stage, has reached the startling conclusion that our belief in the spontaneity (randomness) of mutations is extremely insecure and that there is circumstantial evidence that some bacteria can choose which mutations to produce. See John Cairns, Julie Overbaugh, and Stephan Miller, "The origins of mutants", *Nature* (September 1988), **335**: 142–5; see also "How blind is the watchmaker?", *The Economist*, 24 September 1988.

mother of invention" type of world.[6] Even if in biology this number is zero, as orthodox adherents of the Darwinian synthesis insist, there is no reason why systems in which it is not zero could not be analysed by basically the same tools. The crucial point is that the outcomes of the process cannot, by definition, be foreseen (or else the knowledge cannot be said to be new).[7]

The selective retention concept is also problematic. Unlike Darwinian theory, which regards natural selection as a metaphor, in production techniques the choice is literal and often based on an explicit and conscious weighing of alternatives against one another. Techniques compete with one another just as much as firms do; the difference is that in this model the firms do the selecting rather than being selected themselves. The exact definition of the criteria used are, however, the subject of considerable controversy. The traditional Darwinian criteria — reproductive success and survival (that is, sheer numbers) — seem unsatisfactory criteria in economics. It is not enough for individuals, firms, or even knowledge to survive and to have "offspring" in some sense. Amartya Sen has argued that life should be weighted by some kind of quality index, a concept that is alien to pure Darwinism but essential if some of the Darwinian ideas are to be adapted to the social sciences.[8] A similar weighting scheme can be contemplated here, although at this stage it is not clear what it might look like.

# TECHNOLOGY AND EVOLUTION: THE SEARCH FOR AN ECONOMIC THEORY

An evolutionary framework may be instrumental in answering some important questions. How can we classify inventions into useful groups so that we can

---

[6] In Darwinian (more accurately Weismannian) systems, the variation is believed to be truly random in the sense that a variation that is beneficial is not more likely to appear than one that is not, because variation is believed to be exclusively due to copying errors in the DNA. In any kind of epistemological system, this cannot be true. Inventions will appear stochastically, but the probability of any specific invention emerging is likely to be higher if it is felt to be "necessary". In a later paper, Donald Campbell, the founder of modern evolutionary epistemology, returns to the issue of the statistical properties that variation would have to have (stationarity; orthogonality to environmental conditions) and deems none of these essential. See Donald T. Campbell, "Blind variation and selective retention in creative thought as in other knowledge processes", in Gerard Radnitzky and W.W. Bartley (eds.), *Evolutionary Epistemology, Rationality, and the Sociology of Knowledge* (1960; La Salle, Ill: Open Court, 1987), pp. 91–114; Gerard Radnitzky and W.W. Bartley, "Unjustified variation and selective retention in scientific discovery", in Francisco J. Ayala and Theodosius Dobzhansky (eds.), *Studies in the Philosophy of Biology* (Berkeley: University of California Press, 1974), pp. 139–61.

[7] This parameter was first proposed by Peter Allen. See his "Evolution, innovation and economics", in Giovanni Dosi *et al.* (eds.), *Technical Change and Economic Theory* (London and New York: Pinter, 1988), pp. 95–119.

[8] Sen, "On the Darwinian view".

make historical sense out of them? What is the shape of the normal historical path taken by technological progress and how can we assess its consequences? Why are some societies technologically creative at some points of time but not others? The difficulties that economic theory has had in explaining technology and its history are well known and have been widely and frequently elaborated. In the economics of information and technological change, much of standard microeconomics breaks down. Above all, new information — including all new technological information — is a commodity whose production is subject to increasing returns to scale, public good properties (costly excludability and non-rivalrousness), and does not satisfy the laws of arithmetic. It is hard to believe that such a commodity can be fully analysed simply by relying on the laws of supply and demand.

Following the pioneering work of Nelson and Winter, therefore, economists have increasingly sought to return to evolutionary biology in order to explain technological change.[9] In these models the choice of the unit of analysis is to some extent arbitrary and may vary, depending on the purpose at hand. The more interesting aspect in technological history is without doubt the selection among different species, not among individuals, even if species selection is the outcome of individual selection. The technical knowledge that defines and underlies the technique is isomorphic to the genotype, whereas the actual expression of that knowledge is akin to the phenotype. For the distinction to make sense, we have to assume that the knowledge required to produce an artefact and the artefact itself are separate entities. A "technique" is, however, not a full description of every gear and cog involved in the production process. A good way of thinking what a technique is in this context is to compare it with the genome. The genome does not "describe" every cell in the organism's body; that would require far too much information. Rather, it is a set of *instructions* on how to produce something. These instructions may be very explicit, as in technical handbooks and blueprints, or very implicit, as in Polanyi's tacit knowledge.[10]

---

[9] See Richard R. Nelson and Sidney G. Winter, *An Evolutionary Theory of Economic Change* (Cambridge, Mass.: Belknap Press of Harvard University Press, 1982). For an introduction, see many of the articles in Dosi *et al.*, *Technical Change and Economic Theory*. A more historical approach is taken in Joel Mokyr, *The Lever of Riches. Technological Creativity and Economic Progress* (New York and Oxford: Oxford University Press, 1990), chapter 11; and Basalla, *The Evolution of Technology*.

[10] The set of instructions contains a large number that take the form of if-then and nested if-then statements, so that phenotypes vary according to circumstances. They often use something akin to fuzzy logic: that is, they may well use partial and subjective values. A technique to grow wheat, for instance, contains a set of instructions as to when to plough, when to sow, when to harvest, and it may well contain a statement such as "If it rains a lot, weed twice a week" without necessarily specifying what "a lot" is; hence the fuzzy part of the instructions. The bulk of the genetic material is not actually used in writing these instructions: the knowledge that wheat's Latin name is *Triticum aestivum* and that it yields approximately 3340 calories per kilogram may be quite irrelevant to the actual "instructions" on how to grow it.

The information employed in writing the instructions is a small subset of all information known. The rest of human knowledge plays no role in the ontogeny of production; it is a bit like "junk DNA".

Technological genotypes, when expressed as phenotypes, become observable techniques in use. Like living organisms, they may relate to one another in different ways: independence, rivalry, complementarity, and predation. Knowing these relations is essential if we want to track the ripple effect of technological change on the economy. Most techniques, like most species, have no direct relation with most others, in the sense that a growth in their number does not affect the number of other techniques, their profitability, or any other fitness criterion we care to define. Like seagulls and whales, they may inhabit the same habitat without affecting one another's existence in the reasonable range. Techniques may be *rivalrous* in the sense that an expansion of one affects the fitness criterion of the other negatively. The relation between power loom weaving and hand loom weaving or between the Diesel engine and the steam engine are examples of this. At other times, techniques are *complements*, like symbiotic species: weaving and spinning are symbiotic much like oil refining and Diesel engines. More subtle relations combine elements of both. Steam boats and railroads were in part competitors and in part complements, depending on whether they served the same routes. Crops could compete for the farmer's attention and the consumer's plate while used in a rotation as complements. Finally, techniques broadly defined could have *predatory* or parasitic relations with others, in that increasing *a* favours *b* but increasing *b* harms *a*. Military techniques are parasitic upon civilian production techniques in that fashion, as are other rent-seeking techniques.[11]

The implications of taking an evolutionary approach to the understanding of techniques has profound implications for what economists can and cannot do: it involves an implicit abdication of their ability to make predictions. Their emphasis is, above all, on the description of the historical circumstances and contingencies that led to their creation. The notion of causality is shifted: evolutionary theory explicitly and deliberately stipulates that history is central to the understanding of the form and efficacy of production techniques. Technology, to use a fashionable term, is a-path-dependent phenomenon par excellence because evolution is a non-stationary Markov chain: the present equals the past plus some deviation. Just because it is primarily concerned with non-repeated, unique phenomena in which experimental confirmation or mathematical proof is difficult, does not demote evolution to a "non-science". Testing occurs

---

[11] Parasitic relations are sometimes ambiguous in their impact on the victim (and it is similarly alleged that sometimes soldiers and other rent-seekers *on balance* benefit their preys, through spill-over and other alleged favourable side-effects of military activity). Cf. "Living together: trends in parasitology", *Scientific American* (January 1992), **266**.

indirectly, by examining how well the theory explains the past, just as Darwin said.[12]

The impossibility of making predictions results from the special dynamics of evolution, dominated by learning and scale effects and ridden by contingencies and irreversible choices. Moreover, in technological systems, selection often tends to be frequency-dependent; that is, the survival probability of a "specimen" depends on the *number* of specimens already existing and not just the survival value of their traits. This happens, for instance, when users of a new technology emulate their neighbours' behaviour as a learning-cost saving device, or when there are positive network externalities. Frequency dependency may lead to unpredictable dynamic systems. This vision of history has been tirelessly advocated by Stephen J. Gould and has recently been increasingly stressed by economic historians such as Paul David and economists such as Brian Arthur.

## AN EPISTEMOLOGICAL TAXONOMY OF INVENTIONS

An evolutionary theory of technology needs to deal with historical change, and one of its tasks is to reformulate a scheme to classify inventions. Taxonomies are useful so long as they are not confused with theories. As Gould points out, in a discipline where data come from observation rather than experiment, taxonomy is essential, because rather than a mindless process of allocation of objects into self-evident pigeon-holes, it becomes a theory of causal ordering.[13] Evolutionary theory suggests a classification of inventions by the nature of the new knowledge they create, independently worked out by Christopher Farrell;[14] in what follows, I have adapted his terminology. Farrell's classification distinguishes between three basic types of invention: *mutations*, which create totally new knowledge; *recombinations*, which apply knowledge from one area to another in such a way as to create a genuinely new artifact or method; and *hybrids*, which combine different inventions in a novel way even if the components were all known before.[15]

---

[12] In a letter to Hooker, Darwin wrote that "change of species cannot be directly proved ... the doctrine must sink or swim according as it groups and explains phenomena"; cited in Stephen J. Gould, "Evolution and the triumph of homology, or why history matters", *American Scientist* (January–February 1986), **74**: 65.

[13] Gould, "Evolution and the triumph of homology", p. 86.

[14] Christopher J. Farrell, "Survival of the fittest technologies", *New Scientist*, 6 February 1993.

[15] Farrell's classficiation also contains a fourth group, metamorphs, which he defines (p. 39) as a "mutation or a recombination that causes a dramatic, visible change in the new generation". As I prefer to separate the distinction between the relative magnitude of an invention and its source, I shall not use this concept.

## *Mutations*

When is knowledge so new that we can speak of a mutation? Some scholars tend to take too literally the biblical dogma that there is nothing new under the sun and that all inventions have precursors. The view that "evolution does not produce novelties from scratch, it works on what already exists" is close to a tautology and therefore not very helpful in the context of technology.[16] Of course, mutations are variations on existing material. Most of the genetic material in every mutant is not new. But in the history of technology we can surely recognize new genetic material, and sometimes, albeit infrequently, these mutants represent identifiably new technological "species", isomorphic to speciation. When new departures are sufficiently radical to result in the creation of a new species, we refer to them as *macro*inventions.

Some of these departures may have precedents. Potentially advantageous mutations may have occurred in similar forms before their final emergence but failed for one reason or another. This may happen in natural history as well, but unsuccessful mutations are never preserved for posterity. In the history of technology we know of such events in a few cases.[17] Moreover, the technological macromutation does not necessarily occur over a single generation in one stroke. Quite often macroinventions consist of a discrete number of cumulative steps forward. The steam engine, for instance, required at least three discrete "jumps": the recognition of atmospheric pressure, the realization of the Papin–Newcomen design, and the separate condenser. In addition, innumerable smaller improvements and refinements were necessary before the highly efficient units of the late nineteenth century were produced. At that stage, however, further improvements ("microinventions") started to run into diminishing returns, and a different, if related, species — the internal combustion engine — appeared to compete with it in much the same dynamic pattern.

Mutations are changes in the genotype, that is, information. Most of them are "neutral"; in other words, they do not express themselves in a changed phenotype. Translated into the realm of knowledge, this has the direct and obvious parallel that most advances in knowledge have no direct application to new technology. Only a small fraction of advances in physics, chemistry, mathematics, and biology ever find their way, however indirectly, to production. The

---

[16] François Jacob, "Evolution and tinkering", *Science* (June 1977) **196**: 1164. See also especially Basalla, *The Evolution of Technology*, passim.

[17] The germ theory of disease, one of the truly significant advances of all times and a macromutation if ever there was one, was suggested numerous times before Pasteur and Koch finally drove it home. Pneumatic tyres, to propose another (and more mundane) example, were invented by Robert Thomson in 1845, but did not catch on until they were reinvented by John B. Dunlop in 1888, when their complementarity with other recent inventions assured their success.

parallel tends to become blurry in the finer details, though: in biology, purely phenotypical changes can occur in the appearance or functioning of an organism that are not translated into genotypes and not transmitted to the next generation. Such a distinction may seem otiose in epistemological systems precisely because of the Lamarckian nature of cultural evolution, in which acquired characteristics can be passed on.

## Recombinations

In biological evolution, species are defined by reproductive isolation. Although there are enough exceptions to this rule to be of concern to biologists, disjoint species, in general, cannot mate and so exchange genetic material. Within each species, however, the evolutionary success of biparental reproduction demonstrates the importance in combining and reshuffling different kinds of knowledge. Yet in the theory of evolution, recombination has limited power in explaining the variety and diversity of life, because it can only combine the genetic material of two similar creatures. In production technology, cross-species exchanges occur as a matter of routine because in cultural evolution the information contained in the genotype can be derived from many dissimilar parents. Indeed, the very foundation of "evolutionary creativity" in epistemological systems is that it is able to draw from a diverse gene pool of existing knowledge and combine it in novel ways. In fields as diverse as theoretical economics and chicken breeding, innovations consist, in a great many instances, of applications of existing information to new purposes. The mechanics of technological recombination thus differs from that of living beings: in biology, the genome contains information that can be characterized as a set of linear combinations between the male and the female chromosomes; innovations consist of changing weights. In all knowledge systems, including technological knowledge, no such constraints exist: information can be taken from a large number of sources and *added onto* existing forms. Because technological recombination is multi-parental, the opportunities for innovation through novel combinations of existing knowl-edge are a function of the complexity and diversity of the economy. The number of theoretically possible recombinations rises with the square of the number of techniques in use, and while the actual number of sources is, of course, much smaller, the conclusion remains that this type of innovation is more likely the higher the diversity of the economy.

Consider the famous invention of puddling and rolling by Henry Cort in 1784 combining rolling, in use in the Liège region for the previous two centuries, with the coal-burning reverberatory furnaces in use in both glass-making and crucible steel-making. These ideas had never before been applied in precisely that way to the problem of converting the high-carbon products of blast furnaces into wrought iron. From a purely epistemological point of view, then, Cort's invention was a recombination. Another example is the fusee, a conical device borrowed from crossbow designs by fifteenth-century watchmakers to equalize the changing force of an unwinding mainspring.[18] Watchmaking in its turn was an endless source of ideas for toys, musical boxes, and precision machinery requiring precision-made cogs, springs, and gears. The use of gunpowder, originally intended for entertainment and military purposes for construction and mining purposes, is another example. The indebtedness of Babbage's Analytical Machine to the idea of binary coding of information as used by Jacquard in his famous loom is well known. Less famous is Richard Roberts's multiple spindle machine, which used a Jacquard-type control mechanism for the drilling of rivet holes in the wrought iron plates used in the Britannia tubular bridge. Farrell cites the typewriter as an application of the principle of the piano: the earliest prototypes actually used piano keys.[19] There was a continuous exchange of technical information between the sewing machine industry and the slightly younger bicycle industry in the late nineteenth century. The mechanical ideas embodied in the bicycle then in their turn benefited other industries: the Wright brothers used the ideas of balance and control from bicycle technology in the design of their first prototypes.[20] Even less known is the example of Michael Thonet's invention of bentwood furniture, in which he adapted the technique of bending wood used by cartwrights (through heating and steam) to bend wood, thus creating one of the major industrial innovations to emerge from the Habsburg Empire in the nineteenth century.[21]

## Hybrids

The difference between a hybrid and a recombinant invention is that a hybrid is a combination of two (or more) *artefacts*, rather than the information embedded in them. It is thus a combination of phenotypes rather than a transfer of underlying genetic material from one "organism" to another. A classic example of a

---

[18] Charles Hyde, *Technological Change and the British Iron Industry, 1700–1870* (Princeton, NJ: Princeton University Press, 1977), pp. 88–9; and Lynn White, Jr., *Medieval Religion and Technology* (Berkeley, CA: University of California Press, 1978), p. 308.

[19] Nathan Rosenberg and Walter G. Vincenti, *The Britannia Bridge. The Generation and Diffusion of Technical Knowledge* (Cambridge, Mass.: MIT Press, 1978), p. 39; and Farrell, "Survival of the fittest".

[20] The Westphalian town of Bielefeld, for example, provides a good example of the mutually advantageous horizontal combination of sewing machines and bicycle manufacturers. See Andrew Lohmeier, "Consumer demand and market responses in the German Empire, 1879–1914" (unpublished dissertation, Northwestern University, 1995), ch. 2; and Tom D. Crouch, "Why Wilbur and Orville? Some thoughts on the Wright brothers and the process of invention", in Robert J. Weber and David N. Perkins (eds.), *Inventive Minds. Creativity in Technology* (New York: Oxford University Press, 1992).

[21] Ekaterini Kyriadizou and Martin Pesendorfer, "Bentwood: technological change and industrial organization in the Habsburg furniture industry" (unpublished manuscript, Northwestern University, 1995).

hybrid invention was the automobile: it was not so much a new form as a combination of existing ones in a new form. Comparable with the automobile, if much simpler in design, was the wheelbarrow, a combination of the stretcher and the pushcart which appeared in Europe in the twelfth century. In most cases, hybrid inventions require complementary types of invention that are necessary if the pieces are to work together and the new device is to be made operational. The breastwheel, perfected by John Smeaton in 1759, combined ingeniously the advantages of the overshot and undershot waterwheel and was then supplied by John Rennie with a sliding hatch that allowed the water intake from a continuum of intermediate points between the top (as in an overshot wheel) and the bottom.[22] Hybrid inventions are delayed when one of the artifacts does not fit in with the others. Thus steam power worked well on riverboats when coupled with paddlewheels, but could not properly be adapted to ocean-going vessels before the screw propeller.

## NOVELTY AND SELECTION

Before we can offer some reflexions on the conditions that facilitate technological progress, we need to return to the distinction between "invention" and "innovation". As noted, evolutionary models of technological change consist of mutations, recombinations, or hybrids followed by selection, a duality roughly parallel to the Schumpeterian distinction between invention and implementation-diffusion ("innovation"). Although the distinction between the two has often been confused, an evolutionary model provides a more or less straightforward formal way of looking at it.

The basic idea is to define a set of knowledge comprising everything that an individual knows. The union of these sets constitutes all the knowledge extant in society, delineated by $\Omega$, roughly comparable with what is referred to as the gene pool.[23] Let $g(\Omega)$ be the number of elements in $\Omega$. In addition, we can define *shared* knowledge as the intersection of two sets. When knowledge possessed and used by $i$ is made available and used by $j$, we may say that it has been selected.[24] Technological change is thus composed of *invention*, which increases

---

[22] Joel Mokyr, "Was there a British industrial evolution?", in Joel Mokyr (ed.), *The Vital One. Essays Presented to Jonathan R.T. Hughes* (Greenwich, Ct: JAI Press, 1991), pp. 253–86; and Terry S. Reynolds, *Stronger than a Hundred Men. A History of the Vertical Water Wheel* (Baltimore, Md.: Johns Hopkins University Press, 1983).

[23] The notation should not be misinterpreted to convey a false exactness to these concepts. The definition of "a society" is sufficiently amorphous to render concepts such as "the union of all knowledge in society" somewhat arbitrary.

[24] Selection occurs not just when knowledge has been made available and been acquired by a new individual but when it is being acted upon, that is, used in production.

$\Omega$, and *diffusion* which increases the sum of all intersections, $\Gamma$. The number of elements in $\Gamma$ is $h(\Gamma)$.

We may now point out a number of properties of $\Omega$ and $\Gamma$. First, it is obvious that the maximum size that $h(\Gamma)$ can reach is $\frac{1}{2}n(n-1)g(\Omega)$. This happens when all $n$ individuals know the same thing. Second, it is possible for $\Gamma$ and $\Omega$ to grow independently of each other: if one individual discovers knowledge but shares it with no one, $\Omega$ will grow but $\Gamma$ will not. If existing knowledge diffuses in society without adding to the stock, $\Gamma$ will grow but not $\Omega$. Third, it is clear, however, that the rate of growth of the two will be related. The definitions of $\Omega$ and $\Gamma$ imply that the two are complements and that, with $\Omega$ held constant, $\Gamma$ will run into diminishing returns long before it has reached its theoretical limit. On the other hand, increases in $\Omega$ will make it likely that $\Gamma$ will grow simply because more useful knowledge increases the economic returns on imitation and learning. Finally, roughly speaking, the economic *potential* of an economy depends on $\Omega$ but its *performance* on $\Gamma$. If an individual discovers a new technique, it is unlikely to be of much use unless the knowledge is shared with others. To be sure, a firm that kept a new technique secret could spread some of its benefits without sharing the knowledge. Generally, however, if secrecy is effective, the benefits of technological progress are limited.

The distinction between the generation of new information (invention) and its diffusion to and adoption by additional people may help us to explain the technological creativity of some societies. The point is simply that different factors govern the generation of new technology and its dissemination. The likelihood of innovations will depend on the frequency at which mutations occur, so that we might want to search for the technological equivalent of mutagens. In genetics, the nature of mutagens, such as certain chemicals and radiation, is well known. Intellectual mutagens are still very difficult to identify. The likelihood that recombinations and hybrids will emerge depends on the diversity of the economy and the degree of integration between different areas and sectors.

The equivalents of "mutagens" in the history of technology are those factors that somehow "cause" people to have new technological ideas. It seems obvious that society's tolerance to new ideas plays an important role here. Nothing will inhibit new ideas more than the knowledge that their propounder may be persecuted or ostracized on their account. Conformism is the arch-enemy of progress. Tolerance, it seems, is largely indivisible: in the past, societies that have been relatively tolerant in intellectual and religious matters have also fostered new technological ideas. The degree of conservatism of a system described by cultural evolution, in which information is passed not only from "parent" to "offspring" but also horizontally and diagonally, depends a great deal on the exact mode of transmission.[25] Systems in which information is largely passed on vertically, such as an apprentice-master or father-son teaching system practised by artisans in the European past, tend to be conservative. When information

---

[25] Luigi L. Cavalli-Sforza, "Cultural evolution", *American Zoologist* (1986), **26**: 845–55.

passes from one individual to many, as in schools or books, there is more choice and diversity in the system and it is likelier that innovations will occur, as well as disseminate faster.

The theory of evolution indicates that a successful mutation consists of more than just a potentially advantageous innovation in the genotype. For it to be eventually expressed in the phenotype, certain complementary factors are necessary; otherwise a gene can remain recessive or eventually disappear without ever becoming homozygotic. Similarly, inventions that fail to be accompanied by the necessary complementary factors or attain a minimum critical mass are doomed to failure. One of these complementary factors is the ability to turn blueprints into economic reality: without the presence of skilled craftsmen and high-quality materials, inventions cannot be made practical. Western Europe during the period of the late Renaissance and Baroque, in particular, suffered from endless still-born technological breakthroughs. The British industrial revolution, indeed, is an example not so much of great and pathbreaking new ideas — although these certainly occurred — as of inventions that were practicable at the time, given the skills and materials that were available.

Once new techniques appear, they have to be "selected" or diffused. Van Parijs has argued that the important point is that selection occurs *ex post*, on the consequences of choices, and not *ex ante*, on the choices themselves, which are predetermined by mutations that create variation.[26] Yet surely firms can decide to choose between options on the basis of their expected consequences. Following Nelson and Winter, we may discuss something called a "selection environment", which essentially determines how rapidly advantageous inventions will be picked. The openness, degree of integration, and competitiveness of the system are critical here. The selection of new technology depends on factors that differ from and at times are contradictory to those generating inventions: a strong patent system is thought to encourage invention, but by definition it also slows down the dissemination of the new information generated. On the other hand, many of the factors that seem to encourage the emergence of new ideas, such as tolerance to deviants, a willingness to take risks, and an openness to ideas generated by foreigners also appear to facilitate their diffusion.

The outcomes of the selection process differ from case to case. One possibility is the fixing of the new mutation in the population, driving out the existing technique with which it now competes. History, of course, is full of techniques that became extinct, driven out by more efficient ones.[27] Often, however, the

---

[26] Van Parijs, *Evolutionary Explanation*, p. 52.

[27] The concept of extinction underlines the hazards of the analogies between the history of technology and evolutionary biology. In biology, the information contained in the DNA can survive only in living beings and cannot be brought back once the last being dies, Jurassic Park notwithstanding. In all epistemological systems, knowledge can survive outside its normal carriers. Books, artefacts, museums, and blueprints all make it possible for the information to survive even when extinct and brought back. Yet when the knowledge is to a large extent tacit or deliberately kept secret, it may become impossible to re-create a technology.

new and old techniques co-exist for long periods, with sometimes the new one becoming dominant and the old surviving in a niche, sometimes the reverse. At times, they arrive at a *modus vivendi* and distribute the world between them.[28] Economists often feel intuitively that mechanisms of technological selection are essentially rational and that techniques are selected according to some economic criterion, such as profit maximization, leading essentially to socially optimal outcomes. Such "Panglossian" views are, unfortunately, not consistent with the historical record and have been heavily criticized by economists committed to evolutionary paradigms.[29] Favourable inventions are often fiercely resisted by the system and fail to be selected.

There are two types of resistance to the selection of a new and possibly superior technique. The first is related to "interrelated technological systems" or "network complementarities", as they are known today. It may be impossible to change one component in a system without changing other components because of the way they interact: electrical equipment, trains, software, and farming in open field agriculture all shared the problem of interrelatedness. Yet here, also, the analogy can be pressed too far. In technology — but not in nature — we can invent "gateway" technologies in which the incompatibilities are overcome. These would include, for instance electrical converters from 115 volts to 220 volts or railroad cars with adjustable axes that travelled on different gauges. Positive feedback traps can occur in technological systems, but they tend to be rare in open economies because of competitive pressures from outside.

System externalities have an equivalent in biology known as "structural constraints". Genetic material is transmitted in "packages" and so sticks together. The information transmitted from generation to generation does not consist of independent and separately optimizable pieces. A "little understood principle of

---

[28] An example is the history of engines. Wind power did not drive out water power, and steam did not drive out either one; it divided the world with water power, though wind power gradually became confined to niches. The high-pressure steam engine, similarly, did not drive out stationary Watt engines but coexisted with them. Two-thirds of a century later, the internal combustion engine emerged and for a time coexisted with steam, although in certain territories (e.g. cars and agricultural machinery) it drove it out. When the Diesel engine appeared, it shared the world with the four-stroke (Otto) gasoline engine but gradually drove steam to extinction. The two-stroke engine seems to have survived in little niches of lawnmowers and the less-than-competitive planned factories of the Trabant. Some seemingly good mutations, such as the steam car and the Wankel engine, did not get selected to the point where they could occupy a niche.

[29] Hodgson, *Evolution and Economics*, chapter 13. For a parallel discussion among biologists, see, for example, Stephen J. Gould and Richard C. Lewontin, "The spandrels of San Marco and the panglossian paradigm: a critique of the adaptionist programme", *Proceedings of the Royal Society of London* (1979), **205**: 581–98; and John Dupré (ed.), *The Latest on the Best: Essays on Evolution and Optimality* (Cambridge, Mass.: MIT Press, 1987).

correlated development" (as Darwin called it) implies that certain features develop not because they increase fitness but because they are correlated with other developments. We now know why this is so: genetic linkage causes genes that are located in close proximity on the chromosome to be inherited. At the same time, evolution tends to be localized and cannot change too much at once. As François Jacob put it in a famous paper, evolution does not so much create as tinker: it works with what is available, odds and ends, and much of it therefore involves minor variations on existing structures.[30] Selection could also misfire when a trait leads to "positive feedback traps", that is, selection of a trait because of its success in satisfying the fitness criterion but trapping it at a low level of fitness.[31]

The other factor affecting "selection" of techniques by the selecting agents (firms and individuals) is social resistance to new technologies.[32] Technological progress, no matter how beneficial to society, will almost always harm some individuals whose investment in the physical or human capital of the old technology is reduced in value. These individuals may then get organized, whether in the legal and officially sanctioned organizations of craft guilds or the roving

---

[30] See Jacob, "Evolution", p. 1165. Such "minor" variations, however, can have huge consequences on the phenotype. As Jacob notes, "small changes modifying the distribution in time and space of the same structures are sufficient to affect deeply the form, the functioning and behavior of the final product" (p. 1165). For a similar view, see Stephen J. Gould, "Is a new and general theory of evolution emerging?", *Paleobiology* (1980), 6: 127.

[31] Peter M. Allen and M. Lesser, "Evolutionary human systems: learning, ignorance and subjectivity", in P.P. Saviotti and J.S. Metcalfe (eds.), *Evolutionary Theories of Economic and Technological Change. Present Status and Future Prospects* (London: Harwood, 1991). An example is the peacock's tail, which helps each peacock in the reproductive game and thereby conveys a selective advantage despite the uselessness of the tail in survival-related functions. The same was true for the extinct Irish elk: its enormous antlers gave its bearers a putative advantage in mating, but they were apparently useless as a defensive tool and helped in the demise of the species. One can easily think of products that survive because of their success in marketing and advertising despite demonstrable lower quality.

[32] Resistance to novelty, too, is a property shared by biological and technological systems. In a recent paper, Mayr raises the deep and important question of the stability of phenotypes over millions of years despite steady rates of mutation. His explanation turns around "epistatic interactions" that occur because genes are parts of genetic teams that cohere together. See Ernst Mayr, "Speciational evolution or punctuated equilibria", in Albert Somit and Steven A. Petersen (eds.), *The Dynamics of Evolution* (Ithaca: Cornell University Press, 1989), pp. 34–5. For the parallel argument in a history of technology context, see Joel Mokyr, "Progress and inertia in technological change", in John James and Mark Thomas (eds.), *Capitalism in Context. Essays in Honor of R.M. Hartwell* (Chicago: University of Chicago Press, 1994), pp. 230–56.

bands of Ned Ludd and Captain Swing, and resist the new technology. The outcome of the struggle between the Old and the New depends on the political and legal institutions of the society in question, because it is determined by necessity within non-market settings. Although neither pressure groups nor collective action exist in natural selection, the process of innovation in nature and in technological systems still bears some interesting parallels. As Robert Wesson has pointed out, "the most important competition is not among individuals and their lineages, but between new forms and old. The old must nearly always win, but the few newcomers that score an upset victory carry away the prize of the future". This paragraph, written as a comment on Darwinian evolution, mirrors the one written decades earlier by Schumpeter: "In capitalist reality, as distinguished from its textbook picture it is not [price] competition which counts but the competition from the new commodity, the new technology ... which strikes not at the margins of the profits of the existing firms but at their ... very lives".[33] Yet here too, the analogy should not be pressed too far. The coalitions that resist change are maximizing their narrow utility functions, and by so doing they reduce social welfare. The idea that there is some kind of overreaching welfare function to be maximized is, however, an idea peculiar to the social sciences. There is no equivalent to utility maximization in biological systems, and so the "reactionary" resistance to technological change carries a negative connotation that is absent from natural systems.

## SOME LESSONS

An evolutionary analysis of invention has some important lessons for economics. One is that diversity is a source of creativity and thus of progress. Strange as it may sound to non-economists, this insight is rather alien to economic theory, where it is implicitly believed that variety and diversity are signs of inefficiency.[34] In general, evolutionary theory suggests the reverse: diversity is a crucial source of innovation. Diversity results from independent experimentation. The path-dependent nature of evolutionary processes, in nature as much as in technology or the arts, almost guarantees that historical sequences isolated from one

---

[33] Robert Wesson, *Beyond Natural Selection* (Cambridge, Mass.: MIT Press, 1991), p. 149; and Joseph A. Schumpeter, *Capitalism, Socialism and Democracy* (3rd edn., New York: Harper and Row, 1950), p. 84.

[34] Even in consumption theory, where the idea that diversity is the spice of life has intuitive appeal, it has never had much impact, despite the eloquent and learned plea by an eminent economist. Cf. Tibor Scitovski, *The Joyless Economy* (New York: Oxford University Press, 1977).

another will yield quite different outcomes. The fauna on insulated Island ecologies, for example, tends to be quite different from that on the mainland. Similarly, Chinese, Amerindian, and Western medicine and farming developed in quite different ways. The slow osmosis between the quasi-impervious membranes separating Chinese and Western cultures was a source of enrichment for both, as each took over whatever the other did better. Modern economic history is, in large part, a tale of a technological exchange between highly diverse economic and political systems that exchanged ideas.

Diversity, however, also affects the rate at which inventions occur and their nature. In large part, this follows immediately from our earlier discussion of technological recombination as a source of creativity. In all cultural evolution, the mutual enrichment of systems that have evolved separately is an enormous source of creativity.[35] At the simplest micro-level, this is true simply because, as Adam Smith warned us, excess monotonousness is detrimental to creativity. More fundamentally, the unpredictability or "blindness" of mutations means that there is no way of predicting *a priori* which path leads to the most efficient method of production. Selection chooses techniques because of what they are, not what they might become.[36] This "myopia" means that what seemed at the time a rational choice may turn *ex post* into a bad choice because of unforeseen consequences.[37] But, for the economy to realize that a choice was "bad", it has to observe an existing alternative elsewhere, that is, another economy that has made a different choice. Then the two economies can compare notes, and the one that realizes *ex post* the "error" can switch back, albeit at a cost. Diversity thus prevents the foreclosure of what may eventually become viable alternatives. It prevents being locked into what turns out to be inferior outcomes simply

---

[35] In science, for instance, it is held that all macro-mutations require some inspiration from a different area. In his thoughtful and provocative book on scientific innovation, one of Root-Bernstein's characters notes that "Those who are trained and work entirely within an existing niche cannot make a discovery ... they may very well be adaptive within a narrowly defined region of knowledge and work ... but they will not be capable of evolving beyond the niche". He then adds an important afterthought: "Unlike industry, we must strive for maximum diversity, not maximum uniformity". See Robert Scott Root-Bernstein, *Discovering* (Cambridge, Mass.: Harvard University Press, 1989), p. 375.

[36] Pier Paolo Saviotti and J. Stanley Metcalfe, "Presenting developments and trends in evolutionary economics", in Saviotti and Metcalfe, *Evolutionary Theories of Economic and Technological Change*, p. 12.

[37] Myopia in a way is a misleading term here, because it implies incapacity to see what can be seen. In evolutionary systems, the lack of predictability is inherent: there is no way of foreseeing the pool of opportunities that mutation creates, so that an understanding of the selection process is a poor guide. Past science fiction writers have on the whole done a poor job in predicting future technology.

because nothing else was ever tried.[38] As technologies are adopted by alien societies, they are adjusted and improved, sometimes to the point where the society
of origin has to borrow it back. Furthermore, diversity is conducive to invention
because the adoption of useful techniques may be blocked by conservative
attitudes and reactionary institutions and run into a dead end in one society, but
get a fresh start when transplanted elsewhere. When a technique ceases to be
improved in one society, this does not mean necessarily that it cannot be; it
could be that forces of conservatism and inertia are overcoming those of creativity. Inoculation against smallpox, for example, was practised by some midwives
in Turkey, but was not formally part of medicine and not practised on a wide
scale in the Muslim world. When transplanted to the West in the opening
decades of the eighteenth century, it was much investigated and publicized;
eventually Europeans took to the idea of immunization, improved it vastly by
going from inoculation to vaccination, and were eventually able to take it far
beyond anything the Orient had ever dreamed about: the complete eradication of
the disease. Diversity is costly, however. There is a trade-off between the costs
of non-standardization and the benefits from diversity. Hence it may be an irri-
tant for travellers that the preferred mode of public travel in Europe is the train
whereas in the USA it is the aeroplane, but this permits those in each area to
enjoy the learning effects associated with the other's technique. The lack of
standardization of uniformity in car parts and computer software is a clear-cut
cost, but the technological debt that American manufacturers owe to foreign
brands is such that the inconveniences of multiple brands are fully paid for.
Incompatibility between replacement parts and software that serve identical
functions is costly, but at times can have high pay-offs.[39] Apart from such
learning effects, of course, diversity of standards serves no purpose: nothing is
gained from the British driving on the left and the rest of the Continent driving
on the right, or from the USA's using non-metric weights and measures when the
rest of the world is committed to the metric system.

---

[38] Examples of foreclosed options of this kind are easy to come by. Brian Arthur has suggested steamcars as one technological option that was not chosen; see Arthur, "Competing
technologies". Arguably, had steamcars remained operational enough to enjoy the fine-
tuning implied by constant use, they may have turned out superior to internal combustion
four-stroke engines. Another example is airships. The history of aerospace suggests a struggle between two highly fit "species": fixed-wing aircraft and airships. The struggle between
dirigibles and fixed-wing aircraft in the interwar period can be modelled as a bifurcation, in
which accidental factors were instrumental in the complete victory of one technology over
another. Cf. Price Bradshaw, "The role of technology in the failure of the rigid airship as an
invention" (unpublished dissertation, University of Florida, 1975); L. Ventry and E.M.
Kolesnik, *Airship Saga* (Dorset: Blandford Press, 1982); and Douglass H. Robinson,
*Giants in the Sky* (Seattle: University of Washington Press, 1973).

[39] A similar trade-off exists in the "Tower of Babel" effect of languages. The inconveniences of miscommunication and costs of translation are offset against the mutual enrichment of different languages.

In the evolution of innovation, wastefulness may be a positive thing. For economists, trained from childhood to believe in efficiency, such a statement may be grating. It is, however, a direct corollary of the evolutionary paradigm.[40] A similar reversal of traditional economics is the notion of "error". In many economic models, it is understood that optimizing agents make costly errors, which does not contradict the notions of rationality and efficiency as long as these errors satisfy certain statistical properties to make sure that the agents are not systematic in their errant ways and will correct the error when faced with its consequences. In an evolutionary framework, errors play a different role: they are the source of innovation and creativity.[41] In standard genetics, mutations occur because the genetic material is not copied properly. All creativity is, in Monod's famous formulation, a result of "chance and necessity".[42] In evolutionary dynamics, chance manifests itself in stochastic deviations. In the history of technological change, some inventions have resulted from errors, and what might have seemed like a "deviation" has turned out to be a new departure. Inventions are, by definition, deviations from orthodoxy and convention. The vast majority of them are no more than that, and they remain a source of waste; but the intrinsic unpredictability of the innovation process make this inevitable. Evolutionary models therefore affirm Schumpeter's distinction between static efficient models in the short run and the long-run advantages of inefficient, error-making, but innovative systems.

Evolutionary models also suggest a new way in which economists look at long-term historical dynamics. Positive feedback and interaction effects can produce chaos, as biologists have increasingly stressed.[43] There is a certain indeterminacy about the outcome of some technological processes that has been characterized as "chaotic" by certain writers and compared with the behaviour of non-equilibrium thermodynamics.[44] Some authors have concluded that equilibrium outcomes are irrelevant because technology does not lend itself to a mechanical approach embedded in classical physics and because it displays "open system behaviour". The entire analysis has to shift from the determinate mathematics of equilibrium economics to a more stochastic approach.[45] It has

---

[40] Campbell, "Unjustified variation".

[41] Allen, "Evolution, innovation and economics"; and Hodgson, *Economics and Institutions*, p. 288.

[42] Jacques Monod, *Chance and Necessity. An Essay on the Natural Philosophy of Modern Biology* (New York: Alfred A. Knopf, 1971).

[43] For a summary of some of the literature, see especially Wesson, *Beyond Natural Selection*, pp. 155–8.

[44] Saviotti and Metcalfe, "Present developments", and Gerald Silverberg, "Modelling economic dynamics and technical change: mathematical approaches to self-organization and evolution", in Dosi *et al.*, *Technical Change and Economic Theory*, pp. 531–9.

[45] Allen, "Evolution, innovation and economics" and Malte Faber and John L.R. Proops, "Evolution in biology, physics and economics: a conceptual analysis", in Saviotti and Metcalfe, *Evolutionary Theories of Economic and Technological Change*, pp. 58–87.

even been surmised — by a non-economist — that economies at a higher stage of development have sufficient complexity as to be subject to "chaotic, unpredictable evolution".[46] Evolutionary analysis may retain the insights of equilibrium analysis but give up the notion of uniqueness. The most illuminating metaphor is Sewall Wright's idea of an adaptive landscape with multiple peaks each representing a local optimum. The living world is perhaps the best evidence there is of this: if there is an "optimal" shape of leaf, of flying, or of catching insects, most species are far from it. The very diversity of life and the discreteness of species is proof that evolutionary optimization is invariably local. Technological systems are deterministic only in predicting that the system will reach one of the adaptive peaks, but that it may be a matter of chance *which* one will be picked.

Evolutionary theories of technological change imply that small historical accidents and contingencies may permanently alter the course of events and lead to dramatically different outcomes. The main reason why this occurs is feedback: unlike standard models of exogenous variables affecting endogenous ones, in evolutionary models change in part of the system changes the parameters in another part. In the course of its history, the system passes through bifurcation points leading to certain irreversible choices. Yet at the bifurcation points, the reasons for going one way or another may be quite accidental.[47] Small differences in initial conditions at those points may produce radically different outcomes. Science is based on the notion that *identical* initial conditions lead to identical outcomes. Economic history — not having the luxury of ever specifying exactly identical conditions — has had to settle for the idea that *very similar* initial conditions lead to very similar outcomes. The theory of evolution coupled with the insights of modern chaos theory suggests that this hope may be misplaced.

An evolutionary approach implies that technological change is characterized by fits and starts and that most of the total development that has produced the

---

[46] David Ruelle, *Chance and Chaos* (London: Penguin, 1991), p. 84.

[47] Stephen J. Gould has proposed the metaphor of "rewinding" the tape of history and playing it again. In his view, the replay would be quite different. This is not a new view among biologists, even if it has tended to be a minority view. Alfred Lotka, a mathematical biologist, likened history to a ball rolling along the ridge of a straight watershed, in which a small perturbation was sufficient to determine in which valley the ball would descend. Lotka attributed this idea originally to James Clerk Maxwell, and adds that such instability is typical of systems disposing of funds of "available" energy such as living organisms. He cites Maxwell as conjecturing that "all great results produced by human endeavour depend on taking advantages of these singular states [i.e. bifurcation points] when they occur". Cf. Stephen J. Gould, *Wonderful Life. The Burgess Shale and the Nature of History* (New York: Norton, 1989); and Alfred Lotka, *Elements of Physical Biology* (Baltimore, Md.: Williams & Wilkins, 1925), pp. 407–8.

present has been concentrated in small parts of the past.[48] This interpretation rejects the *natura non facit saltum* idea without necessarily rejecting other aspects of equilibrium analysis. An extreme version of punctuationism is a belief in the sudden emergence of radically different forms. In its purest version, the biological theory known as "saltationism" undoubtedly is contradicted by what was subsequently learned about genetics, namely that large mutations are invariably lethal. All the same, legitimate disputes arise about the viability of "hopeful monsters", as Richard Goldschmidt called them. We do, after all, observe catastrophic behaviour in the natural world as well, even if this is not fully understood.[49] The idea that single mutations could form the basis for new subspecies that would eventually become reproductively isolated has not found much support, despite continuous attempts to resurrect it. Yet at the same time it is found that the great Cambrian explosion that brought about the greatest change in life, the emergence of multicellular creatures, took place in the geologically short time of a few tens of millions of years.

Which of these features apply to the history of technology? It is easy to become overly enthusiastic about the "new mathematics of chaos" and throw overboard all notions of stability and order. The history of technology, very much like natural history, is in fact more characterized by stasis than by change. Perhaps the worst mistake we can make is to project the rapid changes of the past century back into history. The last two centuries of technological history can be compared with the great Cambrian Explosion. If the comparison is accurate, they may be anything but representative.

---

[48] In a telling metaphor, Kenneth Boulding (one of the pioneers of the use of biological analogies in economics) has proposed what he calls the "iceberg effect". Historical change is comparable to icebergs suddenly tipping upside down when slow melting of the immersed lower part leads to a gradual change in the centre of gravity. Kenneth Boulding, *Ecodynamics. A New Theory of Societal Evolution* (Beverley Hills, CA: Sage, 1978), p. 330.

[49] E.g. the freezing and boiling of water, involving sudden jumps as a result of small changes in environment. Statistical mechanics is still unable to provide a coherent explanation of these so-called phase transitions; Ruelle, *Chance and Chaos*, pp. 123–4.

# Medieval Technology and Social Change

# Lynn White's *Medieval Technology and Social Change* After Thirty Years

## *Bert Hall*

Lynn White's *Medieval Technology and Social Change* ranks as one of the most widely read and influential works of historical scholarship in the twentieth century. In my own experience, it is often the only serious work about medieval technology that most historians have read. It was always a controversial work, generally better received by historians of science and technology than by conventional medievalists.[1] *Medieval Technology and Social Change* is a work of the 1950s, originally a series of lectures that White gave at the University of Virginia in 1957; then, delayed at the press, it appeared in print in 1962. The book has now reached a respectable middle age, time for a calm look back at the book and its author, who passed away in 1987.

Lynn White was always prone to look for the small, significant detail, and there are three such details which help us understand how he saw his principal book. First, it is dedicated to Marc Bloch, someone whom White never met and who had been dead for nearly twenty years when the book appeared. Second, there is a quotation from St Jerome in the front matter; roughly translated, it says "Do not condemn as insignificant those things without which great affairs could not take place"[2]. Third, White never brought out a second edition of *Medieval*

---

[1] For a listing of White's scholarly and semi-scholarly publications, see my necrology and bibliography: "Lynn Townsend White, Jr. (1907–1987)", *Technology and Culture* (1989), **30**: 194–213.

[2] Non contemnenda quasi parva sine quibus magna constare non possunt.

*Technology and Social Change*,[3] even though the book became a scholarly "best-seller" shortly after its appearance, and by 1973 had appeared in Italian, German, French, and Spanish. Each of these details is significant.

For many historians of technology, White became an intellectual champion, someone who had finally fitted technology as a critically important element into a broader pattern of historical developments. White's *chef d'oeuvre* won the Pfizer Award from the History of Science Society in the year it appeared, and the Society for the History of Technology awarded him its highest honour, the Leonardo da Vinci Medal, in 1964. He later received SHOT's Dexter Prize for his first collection of essays, *Machina ex deo*, an unusual award for a set of essays and one that probably reflects the reputation of *Medieval Technology and Social Change*.[4] White gathered many additional academic and professional honours, some of them, no doubt, on the strength of the book. He was president of the History of Science Society in 1971–2, of the Medieval Academy of America from 1972–3, and of the American Historical Association in 1973.

On the other hand, *Medieval Technology and Social Change* was never very highly regarded by medievalists themselves. Indeed, the only review of the work anyone now remembers is the vitriolic treatment it received at the hands of Peter Sawyer and R.H. Hilton in *Past and Present*. White's Birmingham critics saw his speculative theses as a "misleadingly adventurist" form of "technical determinism" based on "scanty evidence" leading to "obscure and dubious deductions".[5] (All this from the first paragraph!) White joked about the "tone of high emotion" found in the Sawyer and Hilton piece, but beneath the facade, he was puzzled (and, I think, somewhat hurt) by what he saw as a personal attack. He never bothered to reply to his critics. In some measure, he always felt that to do so would lead him into a defence of *amour propre*. In large part, however, he simply moved on to other issues, and this lowered his level of emotional involvement in the book's theses. It is almost certainly for the same reasons that he never bothered to edit a second edition. He looked on the book, as on much of his work, as a pioneering effort, and he viewed most historical theses as inherently transitory. He expected to become outmoded, and he had no illusions about leaving any single book as a monument. In a profession not without its egocentric *prima donnas*, White was utterly without the kind of vanity that fears

---

[3] The pages of the multi-volume *Dictionary of the Middle Ages* (New York: Scribner, 1982–9) contain what might be regarded as the "last redaction" of *Medieval Technology and Social Change*. White wrote two articles: "Agriculture and nutrition: northern Europe", in (1982) **1**: 89–96 and "Technology, Western", in (1988), **11**: 650–64 (the latter appearing posthumously, of course). One of White's students. Carroll Gillmor, was responsible for the article "European cavalry", in (1983), **3**: 200–8.

[4] Subtitled *Essays in the Dynamism of Western Culture* (Cambridge, Mass.: MIT Press, 1968), subsequently reissued by MIT Press as *Dynamo and Virgin Reconsidered*.

[5] R.H. Hilton and P.H. Sawyer, "Technical determinism: the stirrup and the plough", *Past and Present* (1963), **24** : 90–100.

to be proven wrong. So long as the debate itself advanced the cause of learning, the effort was worth while.

White's disdain for defensive arguments leaves it for us to come to terms with *Medieval Technology and Social Change* now that the dialectic processes of claim and counter-claim have worked themselves out. What were White's intentions, and how did they relate to his own formation as an historian? Where did White learn the history of technology, and how did he reconstruct it to serve the arguments he fashioned? White's primary intention, as I see it, was revealed in the quotation from Jerome (and more explicitly in a host of articles):[6] to invite his colleagues to pay attention to the "small things", the little segments of a larger pattern of life as it existed in the Middle Ages. Likewise, he saw himself as the intellectual offspring of the *Annales* school founded by Marc Bloch and Lucien Febvre in 1929. Such a claim has become such a commonplace among historians that it is necessary to remember how unusual it was in America in the 1930s. Yet White was also the product of a very American approach both to the history of the Middle Ages and to the problem of religion in society. It is in this somewhat unexplored middle ground between the *Annales*, history of technology, and the cultural anthropology of religion that we shall find Lynn White.

## BACKGROUND

*Medieval Technology and Social Change* is above all a piece of *medieval* history; it was written by someone who was trained as a medievalist and worked as one throughout his productive scholarly career. Historians of technology tend to emerge from backgrounds in economic history, technology, or the history of science. But White came to technology from a background in medieval political, institutional, and cultural history. His own autobiographical writings[7] claim that he had decided to become a medieval historian in his first year at Stanford University, i.e. about 1924. Medieval studies had grown greatly at American institutions in the years following the first world war, a phenomenon that owes

---

[6] For a splendid example of White on a small topic, see his "Eilmer of Malmsburry, an eleventh century aviator", *Technology and Culture* (1961), **2**: 97–111. Like many of his articles, this is reprinted in *Medieval Religion and Technology* (Berkeley and Los Angeles; University of California Press, 1978).

[7] "The study of medieval technology, 1924–1974: personal reflections", *Technology and Culture* (1975) **16**: 519–30. A somewhat longer autobiographical piece, written at a slightly earlier date, is "History and horseshoe nails", in L.P. Curtis (ed.), *The Historian's Workshop. Original Essays by Sixteen Historians* (New York: Knopf, 1970), pp. 47–64. White has recently been the subject of a Master's thesis at the University of Oklahoma supervised by Kenneth L. Taylor, in the College of Liberal Studies, University of Oklahoma; see Judith Machen, "Cultural values and the vitality of the West. The mind of Lynn White, Jr." . I am grateful to Mrs Machen for sharing with me a copy of her study.

much to the man White sought to study under, Charles Homer Haskins.[8] Haskins is remembered among historians of science as the author of *Studies in Medieval Science* (1924) or the chapter on science in *The Renaissance of the Twelfth Century* (1927), or even for bringing George Sarton and *Isis* to the Ivy League. Medievalists, on the other hand, remember Haskins as the founder of the Medieval Academy of America and its journal, the redoubtable *Speculum*, or as the French-trained author of *The Normans in European History* (1915), *Norman Institutions* (1918), *The Rise of Universities* (1923), or *Studies in Medieval Culture* (1929). Haskins was more than an influential academic; he was also an advisor to, and apparently a close acquaintance of, Woodrow Wilson, in which role Haskins attended the Paris Peace Conference and played a role in creating the former nation-states of Czechoslovakia and Yugoslavia.[9] The 1920s were a good period for medieval studies in the United States, and White was fortunate in the timing of his career choices.

Before going to Harvard, White spent a year in New York and received a Master's degree from Union Theological Seminary (1928), an interlude that probably reflects the influence of his father, a Presbyterian professor of Christian ethics. (White remained an active Presbyterian for his entire life, and he possessed a nearly professional knowledge of theology.) At Harvard, White's good fortune failed him. After beginning his studies with Haskins in 1929, the mentor suffered an incapacitating stroke in 1931 that left him unable to continue supervising students. White sought Ph.D. supervision from George LaPiana, a specialist on Sicily, an association that left its mark on his dissertation — a history of Norman religious institutions in Sicily. Published in 1938 as *Latin Monasticism in Norman Sicily* by Haskins's Medieval Academy, the thesis was re-issued in 1968.

The transformation of a rather conventional medievalist into a historian of technology is one White himself presented as almost entirely fortuitous. While at work in Sicily in January and February 1933, he learned of the Reichstag fire and Hitler's rise to power. Like many American intellectuals of the day, White was an isolationist, felt that another war was inevitable, and expected the coming conflict to drag on interminably within Europe, cutting off the very life-blood of medieval studies by preventing access to European archives. While still in Sicily, White decided to focus on some field where significant work could be done in the absence of archival materials (without, however, really knowing what that field could be). Later that year, in the fall of 1933, White was a very junior instructor at Princeton. There, while preparing for a course he was assigned, he read Alfred L. Kroeber's *Anthropology* (1923), and he was deeply impressed, he says, by the ability of anthropologists to reconstruct societies that

---

[8] There is no biographical study of Haskins, nor of his chief intellectual descendent, Joseph Strayer (see below). The relevant chapter in Norman Cantor, *Inventing the Middle Ages* (New York: Morrow, 1991), pp 245–86, though controversial in the eyes of many, remains a starting point.

[9] Cantor, *Middle Ages*, pp. 252–3.

lacked written texts, and by the focus on tools, which he credits with calling his attention to technology for the first time. "I have not", he concludes, "been the same since".[10] Reading further, he encountered F.M. Feldhaus's *Die Technik*, the studies of Lefebvre des Noëttes, and, above all, the work he so deeply admired, Marc Bloch's *Les caractères originaux de l'histoire rurale française.*

Looking critically at White's claims, we can see that his choice of anthropological pundits was both adventitious and fateful. Alfred Kroeber was the doyen of American anthropologists, and his textbook was one of the most widely read works in anthropology in its time. Since anthropologists are generally more interested in material culture than historians, it would be natural to conclude that White acquired his interests in technology from Kroeber, and this is pretty clearly the impression that White intended to convey in his autobiographical remarks. Yet Kroeber is not widely regarded among anthropologists as someone who focusses sharply on technology, and certainly not as someone who places tools at the centre of his interpretation of cultures. Students of anthropology's development in this century would classify Kroeber as an ethnologist, a cultural anthropologist and even a "culturologist", an interpreter of culture in the fullest anthropological sense of the word.

Kroeber's work stands in marked contrast with that of someone far more deeply devoted to material culture, to examining technology, and to materialist explanations of culture, Kroeber's professional rival Leslie A. White, for example. Something of Kroeber's approach may be gleaned from a paper he published in 1919. In this paper, "On the principle of order in civilization as exemplified by changes in fashion",[11] Kroeber sought to plot regular curves of change in the ratios of physical dimensions of women's fashions and to argue for a cyclic theory of cultural evolution. He republished his somewhat curious study several times, and reprinted it as late as 1957 in his *Style and Civilization* (Ithaca, NY, 1957). His central focus was always on culture, something he saw, as he said in 1917, as a "superorganic" entity whose primacy and autonomy cannot be explained away by historicism, biological reductionism, or outright racism.[12] It would be wrong, therefore, to conclude that Kroeber gave White only an appreciation for tools. Patently, White also learned from him to look at culture very broadly and appreciate the "little things", the *parva* of St Jerome, within the broadest of interpretive frameworks. White indicates something of this when he states that he was inspired by Kroeber "to apply the methods of cultural anthropology to the Middle Ages".

Marc Bloch's influence and that of the then young *Annales* school are equally visible. If one came to know the *annalistes* from a later perspective, it can be

---

[10] White, "Horseshoe nails", pp. 49–50.

[11] *American Anthropologist* (1919), **21**: 235–63.

[12] "The superorganic", *American Anthropologist* (1917), **19**: 163–213, with discussion on p. 441 ff. I am grateful to Mr David McGee of the University of Toronto for calling these articles to my attention.

surprising to turn to the sulphide-brittle pages of copies of the *Annales d'histoire économique et sociale* from the 1930s to learn how enthusiastically Bloch and Febvre embraced the history of technology. When Henri Pirenne died, in the autumn of 1935, Bloch and Febvre devoted the memorial issue of the *Annales* to "Les techniques, l'histoire et la vie". It was here that Bloch's famous essay on the medieval water-mill appeared,[13] as well as essays by Febvre on the state of the history of technology and desirable approaches to it.[14] Here we see, as if for the first time, calls for an integrated treatment of technology within historical contexts, along with attempts to fulfil this programme through model articles. In the very next issue, Bloch criticized as simplistic and deterministic the work of Richard Lefebvre des Noëttes on horse harness.[15] Much of the driving energy and many of the organizing themes found in White's subsequent work (and, more generally, in the American historiography of technology that emerged later) can be seen in the *Annales* of the 1930s. There are too many points of comparison between White and Bloch to mention. I shall just call attention to how White's penultimate published article unconventionally couples "Agriculture and nutrition" and thereby echoes Bloch's similarly titled piece of 1934, in the *Encyclopédie française*.[16] More significant is White's life-long insistence that medieval attitudes towards technology must be reconstructed from non-literary, even non-verbal, expressions,[17] a stance that recalls the *Annales* school's preoccupation with the notion of *mentalité*.

Kroeber and Bloch — what a heady brew for a conventionally trained twenty-six-year-old in search of a new career track! White's new allegiances led him into conflict with his more orthodox (and more senior) Princeton colleagues. White complains that his faculty colleagues at Princeton were "the most intellectually conservative in my experience, before or since".[18] But what newcomer of his age would not have found his older colleagues a bit hidebound? Perhaps this should be read as the lotus-eater's complaint, the resent-

---

[13] "Avènement et conquêtes du moulin à eau", *Annales d'histoire économique et sociale* (1935), **36**: 538–63.

[14] "Reflexions sur l'histoire des techniques", *Annales d'histoire economique et sociale* (1935), **7**: 531–5. See also Febvre's "Techniques, sciences et marxisme" on pp. 615–23, and "Pour l'histoire des sciences et des techniques" on pp. 646–8.

[15] Marc Bloch, "Les 'inventions' mediévales", *Annales d'histoire économique et sociale* (1935), **7**: 634–43.

[16] Carole Fink, *Marc Bloch. A Life in History* (Cambridge: Cambridge University Press,1989), pp. 184–5.

[17] Most fully expressed in his "The iconography of Temperantia and the virtuousness of technology", in T.K. Rabb and J.E. Siegel (eds.), *Action and Conviction in Early Modern Europe. Essays in Memory of E.H. Harbison* (Princeton, NJ: Princeton University Press, 1969).

[18] White, "Horseshoe nails", p. 50.

ment of the newly enlightened. Certainly as an autodidact in both economic history — to give Marc Bloch's area of expertise its most limited name — and cultural anthropology, White could have anticipated that his attempts to fuse these fields within studies focussing on technology, but without deep allegiances to the methodologies of either field, would lead to friction with the historical profession's conservatives. White later admitted that he produced works that "violate ... [established] presuppositions about the writing of history".[19]

On the face of it, White's reinvention of himself in his twenties was a kind of religious conversion, one that gave him the courage to be unconventional. It allowed him to become quite different from most of his medievalist colleagues, to be broader and more assertive, and to seek grand patterns. It was the conviction that the little things really do matter which permitted him to broach technology despite a lack of formal training in either science, engineering, or mathematics. White's approach to technology as a cultural historian allows great latitude, though it also involves contradictions. In common with Kroeber, White seeks to see technology as a cultural phenomenon. However, he also seems trapped by older views of technology as something grounded in a different rationality, as a logic of engagement with the natural world unmediated by culture in any serious way. For White, different cultures may express themselves through different technologies, but usually only in very simple ways, by "accepting" or "rejecting" an "innovation". There is, I think, a tension in most of White's work between his subject and his methodology, a tension that informs his thought and simultaneously leads to difficulties.

By 1938, White had left Princeton for Stanford; in those pre-McCarthy days, one went to California seeking intellectual freedom. In 1940 he published, as an essay in *Speculum*, a bibliographic survey of medieval technology under the title "Technology and invention in the Middle Ages". The thematic notes White sounds in his survey were to reappear in *Medieval Technology and Social Change* 22 years later. Technology is defined, exactly as we might expect from a cultural historian, as simply "the way people do things". With this declaration, White embraces the realm of the banausic, Jerome's *parva*, discarding any moral criteria and any sense of "good taste" as a guide to what ought to engage the historian's attention. In common with the *annalistes*, White argues that technology does not recognize conventional boundaries of class, geography, or chronology. Going beyond his mentors, he asserts that its natural frame is coterminous with that of the great religions and their associated cultures: Christianity, Islam, Hinduism, Buddhism, and Taoism. White saw technology as the artistic creation of the sub-literate masses of humanity, the "silent majority" (a term he coined and was deeply embarrassed to find used by Richard Nixon, whom he despised). He argued always that technology was a universal and a

---

[19] White, "Horseshoe nails", p. 58, apropos of the reception of *Medieval Technology and Social Change*.

unifying force in human affairs, and he saw its study as an antidote to the tendency towards parochialism which is inherent in most historical studies and which the Nazis had perverted to such malignant ends. White sees technological continuity across the so-called Dark Ages (a thesis much more radical in 1940 than today), and across the sweep of Eurasia, thus involving China with Europe in a vast and complex system of inventions and influences — a vision that led Joseph Needham to recognize White as a kindred spirit.

Even more significant was the keynote White sounds in the closing paragraph of his 1940 essay. Medieval technology is not, he asserts, merely an aspect of economic history, but a "chapter in the conquest of freedom". Technology is "humanitarian", not "rooted in economic necessity"; it is specifically related to "the activist or voluntarist tradition of Western theology". Speaking as a cultural historian, White declares: "It is ideas which make necessity conscious". Technology, especially machinery that saves human labour, is produced by the "implicit theological assumption of the infinite worth of even the most degraded human personality, by an instinctive repugnance towards subjecting any man to a monotonous drudgery which seems less than human". These utterances firmly establish White as a cultural historian, and go a long way to defend him against the charge laid by later uncomprehending critics that he was a "technical determinist": he was, if anything, a "cultural determinist". But it is also through this dogmatic assertion of a linkage between Latin Christianity and Western technology that White's later thought would reach its most extended and controversial thesis.

The underlying thrust of White's thought is apparent even in his earliest essays, and it descends from the programmatic assumptions of his mentors. Simply stated, it is the belief that the Middle Ages make sense, that there is a certain rationality to the medieval order, to the medieval way of doing things. Haskins, along with his protégé, George Sarton, and his intellectual heir, the Princeton medievalist Joseph Strayer, were all part of an effort to explicate the Middle Ages to a generation reared in the fuzzy, romantic Victorian fantasies about the period. They were an important part of what Norman Cantor has rightly labelled the "invention of the Middle Ages", an historiographic movement that, in its way, is one of the quiet triumphs of twentieth-century learning.

The critical task of the early pioneers of modern medieval studies, and especially of the Americans, was to blunt the viewpoint articulated by Jacob Burckhardt, that medieval Europe lived in a "veiled state" of "faith, illusion, and childish prepossession", a state that was only overcome through the superhuman efforts of the Italian Renaissance. The long-running bloodless Balkan feud between medievalists and Renaissance specialists began in the early decades of this century, and it will probably never be settled. For Haskins and his followers, medieval people and institutions were — to use a single word — "rational". Norman administration in Haskins's works, and by extension medieval government in general, was efficient, orderly, and in most other ways comparable with the early modern or the modern state. The Norman conquest of 1066 marked the "modernization" (i.e. the feudalization) of England, and its salvation from the disorder and irrationality of Anglo-Saxon ways. Joseph Strayer was well known for his work on feudalism, that most erratic and seemingly irrational of medieval institutions (or, if you insist, collection of customs), but Strayer's' feudalism is primarily a political and institutional form whose great virtue is that it works, i.e. it is pragmatically "rational".

Strayer's best-known essay, *On the Medieval Origins of the Modern State* (Princeton, 1970), suggests in its very title the other critical element of the Haskins programme: there is no discontinuity between medieval Europe and the present. Modern states, modern ideas, modern institutions, modern habits of thought, modern assumptions, and much more are all descended literally and lineally from a medieval European ancestry. The Renaissance (if it can be said to have existed at all) is a minor movement in arts and letters, in no way comparable with the cataclysmic end of Classical civilizations that gave birth to the Middle Ages. There is therefore no hiatus between medieval and modern times, and "modernity" does not emerge from a revolt against the "medieval". Instead, the modern world was fashioned continuously and predictably (if not always smoothly) from the womb of the Middle Ages. If this medievalist's *credo* were to be reduced to a banner (the medieval equivalent of a bumper sticker), it would surely read, "The Middle Ages made sense, and they *matter*".[20]

These precepts, I believe, help us to make better sense of much of White's work. The main threads of his writing seem to me to amount to a case for the rationality of medieval technology, its meaning within its context, and its capacity to illuminate the modern condition. Like his mentors, White set out to make a Middle Ages that would cohere internally. But this proves to be more difficult in the case of technology than in other fields. For one thing, historians generally come from a conventional, humanistic training that may teach them something about politics, science, and philosophy, but nothing about technology. For another, the continuity between medieval and modern forms of society is evident in a variety of conservative institutions, from the Church to the very nomenclature of officers of the courts: sheriffs, bailiffs, marshals. Technology, on the other hand, claims as its essence a progressivism that leaves the past behind; water wheels, windmills, horse harness, and body armour are now museum pieces, relics of a bygone day. The prejudices of the profession are very real, and White approached with a missionary's zeal the problem of making his colleagues conscious of technology. He had to endure criticism that would have

---

[20] Cantor, *Middle Ages*, p. 285: "All these writers [chiefly White and Robert S. Lopez] were simply applying in extended areas of history of science, economy, and the technology the primacy of rationality that Haskins and most eloquently Strayer perceived in medieval goverment, administration and law".

crushed someone less self-confident than he.[21] In the end, however, White had to abandon the effort as he originally conceived it. When *Medieval Technology and Social Change* appeared in print in 1962, White was already beginning to adopt a slightly different approach to technology, one that moved mainly through cultural history (seen as a species of cultural anthropology) and focussed on the elucidation of attitudes and beliefs about technology. He no longer sought to contextualize the scattered elements that make up technology within the framework of medieval economic, social, or political history.

One further point before turning to cases. Somewhere in his archived correspondence, White refers to himself in a letter to Melvin Kranzberg as "an 'article man' rather than a 'book man'", meaning that he was always more comfortable writing in the terse genre of the article rather than the discursive and expansive style of books. This is evident in his bibliography: apart from *Latin monasticism*, his dissertation, White produced no full-grown scholarly monographs. His writings usually appeared first in journals, then collected inside hard covers. This is also true of the apparent exception, *Medieval Technology and Social Change*. It retains the shape it had in its original incarnation as three lectures, each a separate, article-length study. The entire work occupies only 134 pages of text and footnotes, to which White adds 42 pages of endnotes. With index and illustrations, the entire volume is a scant 200 pages long. His fondness for the briefer format of the article helps to account for White's popularity, particularly in conjunction with his gift for English prose composition. How many people would have read a two- or three-volume *History of Medieval Technology*, especially one written in the customary scholarly style? On the other hand, there is a penalty to pay for this choice; White rarely leaves himself room to consider alternatives, counter-currents, or exceptions. The single-minded preoccupation with a topic, the relentless massing of examples, and the headlong rush to a

---

[21] One example: White had used as an after-dinner speech for several years a little talk outlining the medieval technological ancestry of items usually thought typical of the "Wild West": handguns, barbed wire, playing cards, whisky, etc. This was published in *Speculum* (1965), **40**: 191–202, where it forms a rare example of light-hearted scholarship. A semi-scholarly magazine, *The American West*, reprinted it a year later. White's effort, directed principally towards medievalists, was intended to stake a friendly claim on aspects of an alien territory. Predictably, American specialists in Western history reacted negatively. Owen Ulph of Reed College replied with the requisite ponderous denunciations of White's article as "tendentious", "fantasies", "shaggy generalizations", "sophistries", and worse. Ulph reveals the real source of his irritation when he comments: "The medieval legacy in the United States was, thankfully, almost *nil*." (The thesis of American Exceptionalism leads most Americans to believe that they invented themselves, a fantasy the American historical profession willingly collaborates in maintaining.) Ulph's critique produces the unintentionally farcical effect that always results when academics fail to get the joke. See "The legacy of the American Wild West in medieval scholarship", *The American West* (1966), **3**, no. 4: 50–2 and 88–91.

conclusion make White a "good read". But they may also have led him to over-sell his argument, and contributed to the inevitable critical response.

## THE STIRRUP AND FEUDALISM

The best-known of White's arguments in *Medieval Technology and Social Change* concerns the role of the stirrup in the formation of feudalism. The argument followed a nineteenth-century thesis by Heinrich Brunner.[22] Charles Martel, Brunner argued, had created a cavalry force to repel the Muslim invaders at the Battle of Tours in 732, and had sought to maintain this force by grants of land as fiefs. White modified Brunner's thesis by combining it with archaeological evidence indicating that the stirrup arrived in Francia sometime in the early eighth century. With the stirrup, as he argued, cavalry could fight in quite a new way. The long lance could be held beneath the arm, and blows could be struck home with the combined momentum of horse and rider. "Mounted shock combat" was the exciting prospect made possible by the stirrup, an opportunity of which Charles Martel took free advantage.

White's dramatic explanation of one of medieval history's oldest conundrums were soon criticized,[23] and since then his arguments have been systematically demolished.[24] The consensus of opinion today does not regard the eighth century as a period when cavalry warfare became especially significant, nor, for that matter, is there much agreement on whether cavalry using shock tactics ever played the overwhelming role on the medieval battlefield that White thought it did. Most military historians stress a slow evolution of cavalry equipment and tactics from the Merovingian era onward. Most agree with White's dating of the stirrup's arrival in the Frankish realm — the early eighth century. Most also agree in seeing the stirrup as the first in a series of technical improvements in horsemanship — the high-backed saddle with wrap-around cantle and protective pommel, the double cinch, padded backplates, and breastplate armour being

---

[22] "Der Reiterdienst und die Anfange des Lehnwesens", *Zeitschrift der Savigny-Stiftung für Rechtsgeschichte, Germanische Abteilung* (1887), **8**.

[23] The best review of the controversy is Kelly DeVries, *Medieval Military Technology* (Peterborough, On.: Broadview, 1992), pp. 95–110.

[24] The work of Bernard Bachrach needs special mention here. See his "Charles Martel, mounted shock combat, the stirrup, and feudalism", *Studies in Medieval and Renaissance History* (1970), **7**: 49–75: "Animals and warfare in early medieval Europe", in *L'uomo di fronte al mundo animale nell'alto medioevo. Settimane di studio del centro italiano de studi sull'alto medioevo* (Spoleto,: 1985), **31**: 707–51, and *"Caballus et Caballarius* in medieval warfare", in Howell Chickering and Thomas H. Seiler (eds.), *The Study of Chivalry. Resources and Approaches* (Kalamazoo, Mich.: Medieval Institute Publications, 1988), pp. 173–211.

others — that led to cavalry's apogee. But we now believe that it was only as these improvements became common, towards the late eleventh century, that they led to the widespread practice of holding the lance beneath the arm. The Bayeux Tapestry (c. 1080) shows us a transitional phase in which cavalrymen hold lances in as many as four different positions. By this time, horses themselves began to be bred to more exacting standards, leading to the ultimate medieval war-horse, the destrier, large and sturdy, but fast and nimble as well.[25]

Looking back at the debate, we can see White's urge to make medieval institutions seem "rational". Feudalism is one the Middle Ages' more intractable features. White, like a good Haskins student, shows us the rationality of things, in this case not the rationality of the institution, but of its genesis. Feudalism did not come into being through oversight or weakness, but as the unintended consequence of a bid for military dominance. Carolingian efforts at a short-term gain in military power led to long-term, unforeseen political consequences. There is also the problem of technological determinism. As a cultural historian, White was neither a Marxian nor a materialist and he never saw material conditions as determinants of larger events (though, arguably, his work might have gained in credibility had he been more open to such interpretations). For him, culture is primary; it determines technology. Yet his formulation of the relationship between technology and culture is not entirely clear. For White, Charles Martel's creation of a cavalry force represented an inspired politician's grasping at an opportunity presented by a technological innovation. Commenting on this process, White states:

> As our understanding of the history of technology increases, it becomes clear that a new device merely opens a door; it does not compel one to enter. The acceptance or rejection of an invention, or the extent to which its implications are realized if it is accepted, depends quite as much on the condition of a society, and upon the imagination of its leaders, as upon the nature of the technological item itself.[26]

This is an interesting intermediate position: we might call it constrained choice, or soft determinism. Technologies exist independently of the societies that harbour them, and they create the necessity for choices through their very existence. White also claims, however, that "inventions" sometimes "remain dormant" for long periods of time, only to "awaken" and help "shape" the cultures that use them. This ambivalence, where technology is neither fully socio-cultural nor entirely autonomous, represents an awkward and unresolved tension in much of White's work.

---

[25] For brief versions of the contemporary consensus, see J.F. Guilmartin, "Technology of war", *Encyclopaedia Britannica* (15th edn., revised: New York: Britannica, 1991), vol. 29, pp. 529–47, supplemented by Carroll Gillmor (note 3, above) and Rosemary Ascherl, "The technology of chivalry", in Chickering and Seiler, *Study of Chivalry*, pp. 263–311.

[26] *Medieval Technology and Social Change*, p. 28.

With the perspective of time, we can see that White's identification of the stirrup as an element in horsemanship has become part of a larger colloquy about the nature of medieval warfare, and the debate over the stirrup thesis has moved over into a lively discussion about all aspects of military affairs in the Middle Ages. The entire field, once dismissed as being of no further interest, is fairly buzzing with activity. The very terms of White's argument no longer have much meaning, as feudalism itself has been redefined. White was "wrong". But it is difficult to imagine the current state of the field without his contribution.

## THE PLOUGH AND MEDIEVAL AGRICULTURE

The second-most controversial of White's theses concerns the historical development of medieval agriculture. Essentially, White sought to show the importance of an interrelated series of agrarian innovations: heavy, wheeled ploughs, a three-fold crop rotation system; open-field agriculture practised in common, chiefly on manors; the more systematic cultivation of legumes; and the more efficient use of horses and animal power in general. An economic historian might have chosen to argue that the secular improvement in medieval economic conditions could be traced to such innovations. But White was not an economic historian, and he cast his thesis in more diffuse terms, arguing that the "northward shift in the focus of Europe", and medieval culture's overall "vitality", were ultimately triggered by these changes. It would be difficult to argue against this, but White's dating of these changes and his attempt to relate them in a coherent whole have been criticized. For White, the Frankish realm is once again the locus of changes, and the Carolingian-Ottonian era (roughly 700–1000 AD) the time frame.

Agricultural historians continue to debate the introduction, the rate of diffusion, and the overall impact of these changes. No one denies that such changes took place, and few doubt that they were influential when they did become widespread. But White's attempt to create a consensus has to be judged premature. It is not that a consensus has emerged against White; it is that no consensus has emerged at all. We simply do not fully understand when certain innovations made themselves known. The heavy, three-element plough is a case in point. There is no doubt that the plough with share, coulter, and mouldboard was a superior instrument for turning the heavy alluvial soils of temperate Europe, especially the fertile river-bottoms, but when did such a device appear? Although White was aware of the equivocal character of the evidence, he sought to show that it was imported to Europe through a Slav people late in the sixth century, and that it probably spread westward thereafter. Later research suggests: (1) that the ploughing patterns once thought characteristic of heavy, three-element ploughs can be mimicked by lighter *ard*-type ploughs properly manipulated; (2) that terminological differences may indicate serious variations in types of heavy ploughs; and (3) that the mouldboard plough could have been introduced to Europe at any time from 200 BC onwards.[27] Even the most basic of White's assumptions, his Malthusian belief that population growth results from an improved nutritional regime, has been challenged within the demographic community; the counter-argument, that expanding populations force food production upwards by providing more labour inputs and more demand for food, also has adherents. Seeking general consensus in agrarian history has been likened to herding cats, and whoever wants firm conclusions should probably look elsewhere.

Nevertheless, more recent research in one area towards which White pointed has tended to support and extend his work: horse harness and animal traction. Basing himself in the work of Richard Lefebvre des Noëttes, White pointed to the importance of horseshoes (common by the ninth century) and the horse collar, and the humble whippletree, a lateral bar in harness that permitted tandem teams to be used. The horse collar was appallingly wasteful of an animal's potential in classical antiquity, being maladapted to the horse's anatomy, and the form of the medieval collar permitted dramatic improvements in efficiency. Subsequent research has modified this picture slightly. Spruyette has defended the classical harness system, arguing that by the Roman era, an improved breast-strap system was considerably better for the horse than the older yoke-based system.[28] Equally interesting, from another point of view, is Richard W. Bulliet's work on the camel as a transport animal in the Muslim world.[29] Bulliet argues for the rationality of Islam's dependence on camels and the consequent neglect of the Roman road system in Islamicized areas of the old Roman Empire; he even suggests a camel-based etiology for the medieval horse-collar, citing Tunisia, where camels are used as draft animals. Most significant, however, is the work of John Langdon,[30] who has shown, as an economic historian, how important horse haulage had become in high medieval England. Langdon sees the eleventh century as the period when the horse was introduced to general draught work, and he estimates that by the end of the century, 70 per cent of the energy consumed in English society came from horses and oxen.[31] All of these

---

[27] Bernard Wailes, "Plow and population growth in temperate Europe", in Brian Spooner (ed.), *Population Growth. Anthropological Implications* (Cambridge, Mass.: MIT Press, 1972), pp. 154–79, especially pp. 159–66.

[28] J. Spruyette, *Early Harness Systems*, trans. M. Littauer (London: J.A. Allen, 1983).

[29] R.W. Bullliet, *The Camel and the Wheel* (Cambridge, Mass.: Harvard University Press, 1975).

[30] *Horses, Oxen and Technological Innovation* (Cambridge: Cambridge University Press, 1986): see also his "Horse hauling: a revolution in vehicle transport in the 12th and 13th centuries", *Past and Present* (1984), **103**: 37–66.

[31] Langdon, *Horses*, p. 20. Cited in Joel Mokyr, *The Lever of Riches. Technological Creativity and Economic Progress* (New York: Oxford University Press, 1990), p. 38.

works might well be cited as exactly the sort of investigation White hoped to stimulate through his work.

## POWER MACHINERY

The final chapter in *Medieval Technology and Social Change* is the least controversial, largely because it does not seek to articulate a specific thesis crediting a technological change with some social consequence. Here we can see, with the benefit of hindsight, how White was already shifting his strategy, moving technology towards the higher, but less contested grounds of cultural history. White claims for the later Middle Ages (c. 1000 – c. 1500) the distinction of being "the period of decisive development in the history of the effort to use the forces of nature mechanically for human purposes".[32] There follows a *tour d'horizon* of machine design and power technologies ranging from Hellenistic antiquity to the Baroque, from wind and water power to clocks and fine instruments. This is the largest of *Medieval Technology and Social Change*'s three studies (54 pages of text, plus 17 of endnotes), and certainly the one that has stood the test of time best. There are individual claims, naturally, that have been subject to correction over the years; examples are White's quern and cranked motion argument, for example, or his questioning of the authenticity of sawmills in Ausonius' *Mosella*.[33] But, for the most part, scholarship in this field has built steadily on the foundation White laid.

When James Lea Cate, reviewing the book in *American Scientist*, noted that "for the scholar the longer notes at the end alone are worth the price of admission", he was making a prophetic statement about this chapter in particular. If we shared the habit some other professions practice of tabulating footnote references, there can be no doubt that these pages of White's book would form a dense cluster in the indices, and one that would probably remain fairly constant over the past thirty years. To gain some sense of how the field has emerged since the 1960s, a glance at the relevant annual bibliographies of *Isis* or *Technology and Culture* is proof of the steady increase in this field's activities. A recent survey work seeking to relate technology and economics, Joel Mokyr's *The Lever of Riches*,[34] manifests White's continuing influence in its discussion of medieval technical accomplishments, and likewise in Mokyr's assessment of their significance as well. "By 1500", Mokyr notes, "technology in Europe had

---

[32] *Medieval Technology and Social Change*, p. 79.

[33] D.L. Simms, "Water-driven saws, Ausonius, and the authenticity of the *Mosella*", *Technology and Culture* (1983), **24**: 635–43.

[34] Subtitled *Technological Creativity and Economic Progress*; see note 31, above. Especially relevant are pp. 34–56.

advanced far beyond anything known in antiquity". The economic twist that he places on this increase in knowledge (it was private sector, goal-oriented, "the stuff of Shumpeterian growth") shows that White's synthesis, like the melodic line of a Bach cantata, can play well in a variety of arrangements.

## CONCLUSION

The publication of *Medieval Technology and Social Change* in 1962 marks off the last twenty-five years of Lynn White's life. He remained actively involved in scholarship until shortly before his death; indeed, his single most famous article, "The historical roots of our ecologic crisis", appeared only in 1967. By suggesting that Christianity bears a considerable share of the moral burden for our environmental woes because of its hostile and domineering attitudes towards nature, White roused a storm of protest, but his essay also signalled a turn away from his earlier, more optimistic vision of technology as an outcome of Christian humanitarianism.[35] In this final period — which deserves another full paper, and not just a coda to this one — White became more and more involved in explicating the connections between religious belief and Western technology,[36] and even later with the cultural and technological connexions between Europe and South Asia. These are, obviously, not themes that are at present considered central to medieval history. The great expansion of medieval studies was a phenomenon White helped foster in a variety of ways, but it moved along tracks he had already left behind: political history, ecclesiastical studies, philology, and various literary disciplines.

His work remains a unique contribution, unlikely to be matched in our life-times. In the panoply of medievalists, he remains an "outrider",[37] someone whose work is far more influential in the history of technology than in its native context. Any final judgment on his effort to integrate technology within the web of cultural history remain elusive. His insistence that values derived from

---

[35] The episode is analysed by Elspeth Whitney, "Lynn White, ecotheology and history", *Enviromental Ethics* (Summer 1993), **15**: 151–69. I am grateful to Professor Whitney for providing me with an advance typescript of this work.

[36] Two recent studies that build upon and also criticize White deserve mention here. George Ovitt, *The Restoration of Perfection. Labor and Technology in Medieval Culture* (New Brunswick, NJ: Rutgers University Press, 1987), and Elspeth Whitney, *Paradise Restored. The Mechanical Arts from Antiquity Through the Thirteenth Century [Transactions of the American Philosophical Society*, vol. 80, part 1; Philadelphia, 1990.]

[37] Cantor borrows this concept from the social and natural sciences, where it refers to legitimate data points that for some reason do not fall within the governing paradigm. He uses it as a metaphor for inflential medievalists like Johann Huizinga or M.M. Postan, whose work seemed too "bold, difficult, problematic, or extreme" to become part of the main stream. See *Middle Ages*, p. 376.

religious beliefs are central to the shape and direction of Western technology does not sit well, even today, with engineers (who would pretend to be without values) or with theologians (most of whom appear more intent on absolving religion of any responsibility for the whole of the modern world). For White, it was always sufficient to have instilled some new ideas, or to have aroused dormant thoughts, in his readers. He believed and acted as he preached: that one should work hard and be happy when one's errors find someone to correct them.

Marc Bloch, the man White admired more than any other historian, wrote the words that might sum up White's efforts. With his characteristic simplicity, Bloch published in 1935 a brief comment on a long synthetic article by Alphons Dopsch. Noting that Dopsch's position had received much criticism, Bloch observed that:

> Nothing is more typical of the ordinary development of a science [than criticism]. A system of interpretation, or rather a group of working hypotheses, invariably provokes challenges by the research that it has stimulated. Indeed, the creation of a work of synthesis is rarely ever the achievement of the pioneering scholar.[38]

As heirs to White's legacy, it is up to us to decide what to do with it. The field he pioneered has matured; our tasks are not the same as his. We need, I think, to recontextualize medieval technology, to decide for ourselves what it means. That is as it should be — "the creation of a work of synthesis is rarely ever the achievement of the pioneering scholar". What is important is that we build from the base he has left, that we move forward in directions we choose. Somewhere, I am sure, Lynn White is enjoying the prospect of a wonderfully good argument.

---

[38] "Rien de plus conforme à la marche ordinaire du progrès scientifique. Un système d'interprétation, autrement dit un groupe d'hypothèses de travail, résiste rarement tout entier à l'épreuve des recherches que sa mission était de provoquer. Mais l'oeuvre de mise au point n'est presque jamais dans le destin du premier architecte". See *Annales d'histoire economique et sociale* (1935), **7**: 102. The translation is that of Fink, *Marc Bloch*, p. 152.

# Medieval Technology and the Historians:
## The Evidence for the Mill*

*Richard Holt*

The impact made by Lynn White's *Medieval Technology and Social Change*, published in 1962, was deep and has proved to be lasting.[1] More than any other of White's works, it was this short book — of only 134 pages of text and a further 43 of notes — that was responsible for propagating the belief, now prevalent among historians of technology, that western European society in the Middle Ages was unusually inventive. White portrayed a culture that was eager for change and actively seeking ways of saving human labour, which at different times found itself radically transformed — in both its social structure and its material base — through its choice of new technologies. At the very heart of all White's published writings from 1940 onwards is the conviction that the people of the medieval west possessed considerable technical skills which they were eager to refine and develop, to the extent that the technological developments of the period, and more fundamentally a new mentality of innovation and expansion, were the true forerunners of the industrialization of later centuries.[2]

* I should like to thank Bert Hall, Brian Shoesmith, and everyone else at the "Technological Change" conference who took the trouble to discuss medieval technology with me. My special thanks are due to Sissel Myklebust for her interest and her comments, then and since.

[1] L. White, Jr., *Medieval Technology and Social Change* (Oxford: Oxford University Press, 1962).

[2] L. White, Jr., "Technology and invention in the Middle Ages", *Speculum* (1940), **15**: 141–59, reprinted, with other articles, in L. White, Jr., *Medieval Religion and Technology* (Berkeley CA: University of California Press, 1978); L. White, Jr., "What accelerated technological progress in the Western Middle Ages?" in A.C. Crombie (ed.), *Scientific Change* (London: Heinemann Educational Books, 1963), pp. 272–91; L. White, Jr., "The expansion of technology 500–1000" in C.M. Cipolla (ed.), *The Fontana Economic History of Europe*, vol. 1 (London: Fontana, 1972), pp. 143–74.

Thus, he believed, the prosperity of modern western Europe — indeed, the economic and cultural dominance of the west — had their roots in the ingenuity of medieval men: in agriculture, in warfare, in shipbuilding, and above all in mechanical techniques, and in the harnessing of natural sources of power and the effective exploitation of the horse. Moreover, it seemed to White that the impetus for this technological growth was not economic necessity, which was present in all other societies before and since; it was rather a value-system derived from the Christian church, and specifically from western Catholicism. Only the ideal, rooted in Christian theology, of the value of every human life could give rise to the assumption that it was virtuous to save labour, to relieve others from drudgery. He also contended that medieval society itself owed its essential character to technological change. On the arable plains of northern Europe, shifts in agrarian practice — such as the adoption of the ox-drawn heavy plough and three-course rotation of crops in the earlier Middle Ages, and later a switch to horse-hauling — brought prosperity and economic growth; while the introduction of the stirrup enabled men, for the first time, to fight effectively from horseback. So, according to White, was born the chivalric culture of the Middle Ages, which required the whole structure of feudal society to support the expensive and dubious luxury of the aristocratic mounted warrior.

Not that White had by any means been the first historian to draw attention to the fact of medieval innovation in a number of important spheres of activity. By the 1920s the notion of "the Middle Ages, that epoch of ignorance, stagnation and gloom" had begun to be dispelled, most notably by Charles Homer Haskins's influential exposition of medieval cultural development, at about the same time as Abbott Payson Usher described, for the first time, a range of medieval innovations in technology.[3] Interestingly, while Haskins's concept of a twelfth-century renaissance took root and flourished, little attention was paid at the time by medieval historians to Usher's work; Lewis Mumford's *Technics and Civilization*, published in 1934, certainly did far more to open up the study of medieval science and technology.[4] The same decade also saw the redoubtable Marc Bloch's important, though brief, forays into the history of technology, published in 1935.[5] The attention that Charles Singer's multi-volume *History of Technology*, of the 1950s, paid to a range of aspects of medieval technology, and the sympathetic nature of the treatment, demonstrate just how successful these and other historians had been in overturning the previous low opinion of the medieval capacity to use and adapt the methods and machinery of

---

[3] C.H. Haskins, *The Renaissance of the Twelfth Century* (1927; reprinted New York, 1957), p. vii; A.P. Usher, *A History of Mechanical Inventions* (New York, 1929).

[4] L. Mumford, *Technics and Civilization* (New York, 1934).

[5] M. Bloch, "Avènement et conquêtes du moulin à eau", and "Les 'inventions' médiévales", *Annales d'histoire économique et sociale* (1935), **7**: 538–63, 634–43; trans. by J.E. Anderson in M. Bloch, *Land and Work in Medieval Europe* (London: Routledge and Kegan Paul, 1967), pp. 136–68, 169–85.

production.[6] White's contribution, however, stands out from those of his predecessors: only with him do we see a general theory of medieval social development that attributed a major role to technology. Not only did he regard the social apparatus of feudalism as the logical outcome of technological change; industrial capitalism, too, he saw as having its origins in the mechanized production that he claimed was widespread by the fifteenth century at the latest.[7]

The great strength of *Medieval Technology and Social Change* was that it presented, in an attractive and popular way, what is after all a simple theory that is easily grasped by the non-historian. White's enthusiasm for the people of the Middle Ages and their culture is apparent on every page, and is not the least appealing aspect of the book. But the work is shot through with over-simplification, with a progression of false connexions between cause and effect, and with evidence presented selectively to fit in with his own pre-conceived ideas. Few of White's fellow-medievalists have ever subscribed to his interpretation of medieval society, and indeed an unfortunate result of his book was that it led many other historians to view the study of medieval technology with some suspicion. The formidable critiques that began to appear soon after the book's publication have concentrated primarily on his theories of the origins of feudalism, on his use of evidence and his understanding of the early medieval economy and society, and on the inspirational role he assigned to the church.[8] But other aspects of his thesis which have escaped criticism are just as much in need of it: what he said about the medieval drive towards the exploitation of natural sources of power, for example, is crucial to his theories of mechanization and economic development. It is, therefore, important to show that his account of the medieval mill in its various forms is inaccurate and misleading.

## THE WATER-CORNMILL

White took his account of the introduction and the spread of the watermill throughout Europe mainly from Marc Bloch's published work. To that he added his own research into the introduction of the windmill, and drew together enough references to the application of water power to industrial processes other

---

[6] C. Singer, E.J. Holmyard, A.R. Hall, and T.I. Williams (eds.), *A History of Technology* (8 vols., Oxford: Clarendon Press, 1954–84), especially vol. 2 (1956).

[7] White, *Medieval Technology and Social Change*, p. v.

[8] R.H. Hilton and P. Sawyer, "Technical determinism: the stirrup and the plough", *Past and Present* (1963), **24**: 90–100; D. Bullough, "*Europae Pater*: Charlemagne and his achievement in the light of recent scholarship", *English Historical Review* (1970), **85**: 59–105; B.S. Bachrach, "Charles Martel, mounted shock combat, the stirrup and feudalism", *Studies in Medieval and Renaissance History* (1970), **7**: 47–75; George Ovitt, Jr., *The Restoration of Perfection. Labor and Technology in Medieval Culture* (New Brunswick, NJ: Rutgers University Press, 1987).

than cornmilling to convince himself that medieval people were engaged in a restless search for new ways of using water power. He saw the harnessing of the tides to drive mills as an example of this attitude, of medieval man's restless urge to exploit natural power resources. And although the watermill was indisputably an invention of the Ancient World, White followed Bloch in his belief that it should be regarded as essentially a medieval innovation.[9] But the persistent view that the Romans hardly used the watermill is no longer sustainable. The belief arose in the first place from an over-reliance on the fact that literary references to the mill are severely limited, and on the knowledge that animal-powered or slave-powered mills remained in common use during the later Roman centuries. What the much greater, and increasing, body of archaeological evidence shows, however, is that the period during which the mill was becoming widely accepted throughout Europe was not the early Middle Ages as Bloch believed, but the second and third centuries, or even before.[10] We should not be surprised that White, writing in the early 1960s, was unfamiliar with the archaeological evidence, although some of it was already in print; it is unfortunate, though, that the Bloch-White position should more recently have been re-affirmed, and in the most positive of terms.[11]

White's presentation of the Middle Ages as a particularly innovative period depends in large part on his painting as vivid a contrast as possible with what he saw as the stagnation of Antiquity. And while the Ancient World was not an obviously innovative period,[12] it has now been argued persuasively that the extent and nature of this technological stagnation have been misunderstood.[13] If archaeology indicates that the watermill was common in the late Empire, it has also shown that the crank arm was at least known to the Romans — White was convinced that it was an invention of the ninth century.[14] A convincing case can also be made out in support of the view, rejected by White, that the poet Ausonius was indeed referring explicitly to stone being cut by water-driven saws

---

[9] White, *Medieval Technology and Social Change*, pp. 80–9.

[10] Ö. Wikander, *Exploitation of Water-power or Technological Stagnation? A Reappraisal of the Productive Forces in the Roman Empire* (Scripta Minora Regiae Societatis Humaniorum Litterarum Lundensis, Lund, 1984); Ö. Wikander, "Archaeological evidence for early watermills: an interim report", *History of Technology* (1985), **10**: 151–79.

[11] T.S. Reynolds, *Stronger than a Hundred Men. A History of the Vertical Water Wheel* (Baltimore, Md: Johns Hopkins University Press, 1983), pp. 48–51

[12] M.I. Finley, "Technical innovation and economic progress in the Ancient World", *Economic History Review* (1965), 2nd. ser. **18**: 29–45.

[13] K. Greene, "Perspectives on Roman technology", *Oxford Journal of Archaeology* (1990), **9**: 209–19.

[14] J.F. Healey, *Mining and Metallurgy in the Greek and Roman World* (London: Thames & Hudson, 1978), p. 95.

during the fourth century.[15] Hence the contrast drawn by White between the mentality of late Antiquity and that of the early medieval centuries seems no longer so obvious, as the grounds on which it rested at last undergo drastic revision.

The watermill, then, was already widely used throughout western Europe as the Middle Ages dawned. On the Continent there is scattered evidence for its use through the sixth century; more importantly, we have the evidence of the Irish mills, now reliably and precisely dated by dendrochronology.[16] The remains of over thirty mills from the tenth century and before have been identified, the earliest so far being the Little Island horizontal-wheeled mill built in 630, and its vertical-wheeled neighbour of about a decade later. And the Irish evidence for intensive use of the watermill so early in the Middle Ages, it needs to be pointed out, also has implications for the method used by White and other historians of medieval technology: that is, the conflation of scattered and often imprecise references to mills, from different countries and different centuries, upon which firm theories of technological development and diffusion are erected. Clearly the concentration in Ireland of early medieval mill-remains — by far the greatest in Europe — is outstanding; equally clearly, it has no intrinsic significance and certainly does not imply that the Irish people of that period necessarily made greater use of the mill than anyone else in Europe. Instead, it reflects the high level of preservation of ancient timbers in waterlogged peat, and the extensive digging of peat for fuel in modern Ireland.

We must not lose sight of the fact, therefore, that the evidence we have for milling in other parts of Europe in the early Middle Ages may be unrepresentative. Taking the case of England, for example: there is no compelling evidence that the watermill was an essential feature of the economy. There are some 49 charter references to mills, from 762 onwards, and there are the excavated remains of precisely two mills — one a vertical-wheeled mill of the late seventh century that was subsequently replaced by a horizontal-wheeled mill, and the other a horizontal-wheeled mill of the ninth century. Both, it would seem, were associated with royal palaces.[17] But is this a misleading impression, that watermills were uncommon in early medieval England? When the charter references are mapped, the result does not show the distribution of mills so much as the geographical distribution of surviving charters.[18] Fortunately,

---

[15] D.L. Simms, "Water-driven saws, Ausonius, and the authenticity of the Mosella", *Technology and Culture* (1983), **24**: 635–43.

[16] C. Rynne, "The introduction of the vertical watermill into Ireland: some recent archaeological evidence", *Medieval Archaeology* (1989), **33**: 21–31.

[17] P. Rahtz and R. Meeson, *An Anglo-Saxon Watermill at Tamworth* [Council for British Archaeology Research Report 83] (London, 1992), p. 156 and *passim*.

[18] D. Hill, *An Atlas of Anglo-Saxon England* (Oxford: Blackwell, 1981), p. 114.

moreover, the historian of early medieval England has Domesday Book to refer to, the unique national survey compiled only twenty years after the political demise of Anglo-Saxon society, which establishes — reliably, we believe — that eleventh-century England had more than 6000 mills. At an average of two for each settlement, this was surely a situation that had only come about over a considerable period of time.[19] The extraordinary fact that, to date, not a single one of those mills has been positively identified by the archaeologist only serves to emphasize how precious is the evidence of the Domesday survey. Without the lucky survival of this remarkable source, historians would surely argue that the mill was a rarity in Anglo-Saxon England and that the ubiquitous manorial mill recorded from the twelfth century onwards was essentially an innovation of the Norman conquerors. As it happens, we know this was not the case; and the frequency with which early mills have been found in Ireland leads to the suspicion that even in the seventh and eighth centuries they may have been a familiar sight in England. We can conclude that Bloch and White were quite mistaken in believing that mills were a rarity in Europe until the tenth and eleventh centuries; the lack of evidence for so much of Europe, though, still prevents any more precise conclusions concerning the numbers and use of the mill in the early Middle Ages.

The unreliability of fragmented evidence is a far greater problem than historians of medieval technology have sometimes acknowledged. An illustration of how precariously based, in fact, received and accepted theories surrounding the medieval mill can be, is the case of the tidemill. For White, this was a deeply significant innovation of the eleventh century: it was "the symptom of a new attitude which was to alter the whole pattern of human life".[20] Jean Gimpel, in equally extravagant terms, has claimed the tidemill to have been "typical of the medieval urge to discover new sources of energy".[21] But the technology of harnessing tidal power had been available to western Europe for centuries, and was probably not a medieval innovation at all. It has now been shown that there were European tidemills in the 630s, for the Little Island mills in Ireland were powered by the tidal waters of Cork harbour; and while this is an isolated example from which it would be rash to deduce too much, it is, incidentally, evidence of incomparably superior quality to that for eleventh-century tidemills upon which so much reliance has been placed.[22] Whether or not the mill at Dover harbour in 1086 was a tidemill we can never be sure, because we know so

---

[19] H.C. Darby, *Domesday England* (Cambridge: Cambridge University Press, 1977), p. 361.

[20] White, *Medieval Technology and Social Change*, p. 85.

[21] J. Gimpel, *The Medieval Machine* (London, 1977; repr. 1979), p. 36.

[22] C.B. Rynne, "The archaeology and technology of the horizontal-wheeled watermill, with special reference to Ireland" (unpublished Ph.D. thesis, University College, Cork, 1988), p. 244.

little about it; the tidemills at Venice in the virtually tideless Adriatic, for which White cited only a nineteenth-century secondary source, do not seem very promising.[23] There is really no reason to believe that the tidemill was an innovation of the Middle Ages, especially as we can now see that the technology — which, after all, is a technology of water control and not of mechanism — was known and exploited by a barbarian society of the seventh century. There is as yet no evidence that the tidemill was known to Antiquity, but the likelihood of that must now be apparent.

# INDUSTRIAL WATERMILLS

A theme that White developed was the application of water power to a variety of tasks, with what he saw as important consequences for manufacturing industry. This, more than any other innovation, has inspired those historians who have followed his approach to medieval technology. Again, we see White falling back on the technique of collecting a thin scatter of references from all over Europe, upon which he then rested his arguments. There is a serious fallacy in this method, beyond the objection that the sample is likely to be highly unrepresentative. Fulling mills, for instance, are recorded in different countries and different centuries, and historians seem always to assume that they represent the same phenomenon and are part of the same technological culture. Yet, in a society as diverse as that of medieval Europe, we need to be aware that these might well have been, and probably were, different machines, derived from different sources and, most importantly, operated under different social circumstances and different conditions of production. An excellent example of such technological diversity is the way in which the horizontal mill functioned side by side with the vertical mill in several parts of Europe until the present century, which ought to stand as a warning against regarding medieval, or early modern, society as being technologically homogeneous.

That being said, White did nevertheless identify the fact that during the Middle Ages water power was applied to a number of different machines, and that at one time or another the mechanization of a range of processes was thought feasible — in particular fulling, metal-working, grinding bark, or sawing wood. But he never asked of his evidence, either in *Medieval Technology and Social Change* or subsequently, the questions a historian might be expected to ask. Presumably he did not consider it important to inquire how often these industrial mills were in fact built, and how far they really impacted upon industrial production, although the lack of this analytical study means that his description of medieval mechanization is of limited practical use. He accepted without reservation the published works of others, such as Bertrand Gille; the

---

[23] White, *Medieval Technology and Social Change*, p. 85.

only case-study of an industrial mill that he could cite was Eleanora Carus-Wilson's research into the English fulling mill, in which she had asserted that the introduction of powered fulling had had the effect of moving the focus of the textile industry from the established centres in the south and east of England to the north and west, where water power was more plentiful.[24] Obviously that conclusion suited White, and nowhere did he acknowledge that medieval historians — who knew that the populous towns of lowland England retained a substantial cloth industry — were always suspicious of this supposed impact of mechanization. The fact that Carus-Wilson's arguments were convincingly refuted by 1965 did not stop White from repeating the claim as late as 1972, or Terry Reynolds from doing so — in exaggerated form — even more recently.[25]

To assess the real significance of these industrial mills, we must examine the evidence in a fundamentally different way from White. Only by collecting evidence for mills of all kinds, and more importantly through extensive research into medieval industry, can we ever arrive at a realistic assessment of the impact water power had on production. Recent surveys of English industries in the Middle Ages, which synthesize quantities of both documentary and archaeological evidence, leave no doubt that throughout the period the production of raw materials and finished goods remained labour-intensive and that virtually no processes benefited from the application of water power.[26] Any survey of the mills recorded in the vast body of surviving manorial documentation will show that metal-working mills, whether they were used in the initial processes of production of iron or other metals or for sharpening edged tools, were very rare, and that mills for grinding the oak bark used by tanners were even rarer; sawmills were unknown in England, as timber was invariably sawn by hand with a two-man saw over a pit. Even the relatively successful fulling mill always returned a profit significantly lower than that returned by cornmills, and was therefore built only where there was a surplus of water power: far from being an exciting new investment opportunity for the lords who controlled water resources, fulling mills represented a decidedly second-best use of power that most lords, particularly in prosperous lowland England, rejected.[27]

---

[24] Ibid., p. 88, n. 7; E.M. Carus-Wilson, "An industrial revolution of the thirteenth century", *Economic History Review* (1941), **11**: 39–60, reprinted in E.W. Carus-Wilson (ed.), *Essays in Economic History* (London, 1954), vol. 1, pp. 41–60.

[25] E. Miller, "The fortunes of the English textile industry during the thirteenth century", *Economic History Review* (1965), 2nd. ser. **18**: 64–82; A.R. Bridbury, *Medieval English Clothmaking. An Economic Survey* (London, 1982), pp. 1–36; White, *Expansion of Technology*, p. 156; T.S. Reynolds, "Medieval roots of the industrial revolution", *Scientific American* (1984), **251**: 108–16, on p. 112.

[26] D.W. Crossley (ed.), *Medieval Industry* [CBA Research Report 40] (London, 1981); J. Blair and N. Ramsay (eds.), *English Medieval Industries. Craftsmen, Techniques, Products* (London: Hambledon, 1991).

[27] R. Holt, *The Mills of Medieval England* (Oxford: Blackwell, 1988), pp. 148–58.

That is far from being the final word on medieval industrial mills. Their impact in different locations may well have been more significant than the English evidence suggests. But until that has been demonstrated, none of the assertions of the contribution that water power made to medieval industry should be taken on trust. At best, they are likely to have been exaggerated; at worst, they may be completely unreliable. To take just one example: what White called the "beer mill", which was important to his theory of medieval mechanization in as much as he presented it as the earliest industrial mill, recorded in Picardy in 861.[28] His description of what this mill might have done was misleading: he said it produced mash for beer, directly implying that it somehow mixed the crushed malt with hot water to produce the mash or wort ready for fermentation. He went on to claim that "there is some reason to believe that a new machine was involved: a series of vertical stamps activated by cams on the axle of the water wheel", although he did not explain that his source was a charter conveying a combined mill and brewery which had nothing more to say about either of them.[29] The introduction of the stamping mechanism was an important event; this charter, unfortunately, tells us nothing about when that might have been. It is reasonable to assume that the mill ground malt to be brewed, but there is plenty of evidence from later centuries to show that that was commonly done in an ordinary cornmill and required no modification of the existing machinery at all.

## THE WINDMILL

The presentation of evidence in a selective way is to be seen, once again, in White's account of the introduction of the windmill.[30] His enthusiasm for the new technology again led him to attach too much importance to the benefits of an innovation, and therefore to misinterpret the circumstances of its introduction. That is especially unfortunate, because of all medieval innovations the windmill is at once the best recorded and among the most interesting: it represented a significant addition to the existing range of sources of power, and it is a clear case of a major machine developed almost *de novo* to fulfil an already defined task — that of milling corn where water power was lacking. So an understanding of how the windmill originated and was accepted ought to provide insights into the dynamic forces conditioning technological innovation in medieval society. Windmills were genuinely an invention of the medieval west, and again the superior English evidence is crucial to a study of this first attempt to escape the geographical limitations of water power. We need to examine that

---

[28] White, "Expansion of technology", p. 155

[29] G. Tessier (ed.), *Actes de Charles le Chauve* (Paris, 1952), vol. 2, p. 225.

[30] White, *Medieval Technology and Social Change*, pp. 85–8.

evidence in some detail, therefore, before we can assess White's contribution to the discussion.

Domesday Book of 1086 demonstrates that population and exploited water resources did not always coincide, so that many communities in dry or low-lying districts of England were not served by convenient mills.[31] That was not a new situation, and why it was only in 1185 that the first windmills were recorded, when a need for such an alternative to water power must have been felt centuries before, is an important question still requiring an answer. It may be, however, that it was not until the twelfth century that carpenters developed the necessary expertise — perhaps through recent developments in military architecture or through a new sophistication of domestic timber-framing techniques — to construct the heavy post mills that had to be stable while also being able to rotate easily to keep their sails turned into the wind.

More than twenty English windmills are recorded and firmly dated for the period between 1185 and 1200.[32] There is no reliable reference to a windmill from before 1185; none of the handful of supposedly earlier windmills proposed by Edward Kealey has any credibility.[33] We can identify with some confidence, therefore, the 1180s and 1190s as the period of the windmill's introduction to England. During the years up to the 1220s knowledge of the windmill was spreading. But perhaps its ability to return a profit to its builders was still in doubt, for although windmills were by then to be found all over eastern England, from north to south, and in the Midlands, they were still relatively rare. Then, however, came a half-century in which ecclesiastical and secular lords were building windmills at an impressive rate: the Bishop of Ely's estates distributed throughout East Anglia had four windmills in 1222, and thirty-two by 1251: an eight-fold increase in less than thirty years, which is echoed in the evidence from other great estates in this predominantly dry region.[34] Significantly, we do not observe the same phenomenon in parts of England well served by water power. By the 1280s, there were hundreds and perhaps thousands of windmills in the eastern counties, where more corn was now ground by windpower than by water; in East Anglia two-thirds, and in some districts even three-quarters, of the mills were windmills. Meanwhile, Oxfordshire had just four windmills, according to the Hundred Rolls survey of the county, which also recorded 144 watermills, and windmills remained quite rare throughout most of wetter, more

---

[31] A. Farley (ed.), *Domesday Book* (London, 1783).

[32] Holt, *Mills of Medieval England*, pp. 171–5.

[33] E.J. Kealey, *Harvesting . the Air. Windmill Pioneers in Twelfth-Century England* (Woodbridge: Boydell, 1987). For a critique of this book, see R. Holt, "Milling technology in the Middle Ages: the direction of recent research", *Industrial Archaeology Review* (1990), **13**: 50–8, on p. 54.

[34] British Library, Cotton MS Tib. Bii, fols. 86–241; Cotton MS Claud. Cxi, fols. 25–312.

hilly western England.[35] This was to remain, in effect, the distribution pattern of windmills for centuries.

By 1300, the number of windmills had effectively ceased to grow, further confirming the essential status of windpower as a substitute for water power, rather than as a desirable motive power in its own right. The windmill had been introduced to meet an identified need for cornmilling resources that water power had not been able to meet, and no more windmills were built once that need was met. Moreover, when the population began to fall after 1350, it was windmills, disproportionately, that disappeared. After 1400, when lords found their milling revenues falling, probably as a result of a weakening of their ability to compel their tenants to use their mills, it was again most often windmills — as more marginal enterprises — that fell into disrepair and were abandoned.[36] The rents that could be charged for windmills were almost always lower than those of watermills, and they had consistently proved themselves more expensive to maintain. By 1450, windmills must have been scarcer than at any time since 1230.

White insisted that what he represented as the rapid spread of the windmill was "fundamental to our understanding of the technological dynamism of that era". But he made no systematic attempt to investigate the true extent of the windmill's impact, and exaggerated what evidence he had found. Three times he cited an unnamed chronicler who in about 1322 had written, apparently, that a major cause of deforestation in England was the demand for vanes for windmill sails, or sailyards — an unquantifiable piece of evidence, but illuminating as a contemporary judgement and a sign to White that here was "an age eagerly exploiting mechanical power".[37] But the evidence demonstrates nothing of the kind. The reference is to be found in the register of Pipewell Abbey, in Northamptonshire, and it comes in a lengthy account of how the abbey had happened to lose all of its valuable woodland during the two centuries since its foundation.[38] The greater part, it was said, had been cleared for agriculture; other substantial areas had been sold off for short-term gain by unscrupulous abbots, and much had been commandeered by over-powerful barons. The abbey's servants had sold timber illegally, and the abbey itself had taken too much wood for its own fuel. Great quantities of trees had been felled for building and repairing churches and manor houses, the abbey's granges and barns, and its watermills and its windmills. Indeed, it was reported that "how

---

[35] W. Illingworth and J. Caley (eds.), *Rotuli Hundredorum* (Record Commission, London, 1818), vol. 2, pp. 688–817; Holt, *Mills of Medieval England*, pp. 17–35, 171–7.

[36] Holt, *Mills of Medieval England*, pp. 159–70.

[37] L. White, Jr., "The medieval roots of modern technology and science", in L. White, Jr., *Medieval Religion and Technology*, p. 81; White, "What accelerated technological progress?", p. 273; White, "Expansion of technology", p. 157.

[38] W. Dugdale and R. Dodsworth (eds.), *Monasticon Anglicanum* (London, 1682), vol. 1, pp. 815–17.

many windmill yards were given away in the times of various abbots no-one knows, not even God to whom everything is apparent". Taken in its context, the statement lacks the significance that White gave it, and it certainly does not imply the enthusiasm for mechanization that he inferred. As evidence, it has little to add to what we learn from manorial sources: that for a short period, from the 1220s until 1300, a significant part of the aristocracy was prepared to build and maintain windmills. This was because they promised a reasonable return on investment, as a way of tapping the profits of milling which had hitherto been monopolized by those lords who controlled water resources.

If, like White, we are to draw general conclusions about medieval attitudes to innovation from the case-study of the windmill, we are likely to deduce that medieval people had no interest in power technology for its own sake. It was not any positive enthusiasm for windpower that was crucial to the spread of the windmill, but rather its value as an investment in areas where competition was weak. Those wealthy, powerful people able to take the decision to build a windmill did so when they saw an obvious advantage to themselves — a more prosaic conclusion than White's, but more in accordance with the evidence.

Close examination of the varied aspects of what has been called the "medieval power revolution", therefore, leads to the conclusion that the term "revolution" is inappropriate.[39] Medieval people were not responsible for the introduction of the cornmill, and the tidemill, too, was possibly an invention of the Ancient World. On present evidence, it can be argued that the medieval period saw a slow, but uneven, evolution in the exploitation of natural sources of power, in as much as there was a broadening of the ways in which water power was used, and the windmill — an important technological breakthrough — was invented to fill a specific need. But the fundamental issue of how productive these mills were, or how their productive capacity at different times related to that of the labour force, remains obscure.

## THE CONTEXT OF TECHNOLOGICAL CHANGE

It is beyond dispute that the Middle Ages was a period of considerable change. Medieval society remained essentially agrarian, dominated by a landowning aristocracy, but throughout western Europe, particularly after 1000, there occurred an expansion of population and settlement, an increasing volume of trade and the commercialization of agriculture, the growth of towns, and the ever-increasing show of aristocratic wealth. And everywhere, within that process of change and expansion, it is possible to identify innovation and development, although these

---

[39] An expression used, for instance, by N.A.F. Smith, "Water power", *History Today* (March 1980), **30**: 37–41, on p. 38.

occurred at an uneven rate. A further criticism to be made of all of White's work is that although he showed an awareness of this expansion, and touched on it in his account of innovations in agriculture, he nevertheless failed to relate it to techno-logical developments in the constructive, informative way that would seem to be called for. Surely the context in which to consider the application of water power to industrial processes is the trend towards commercialization that gave parts of Europe an urban or non-agricultural sector exceeding twenty per cent of the popula-tion by 1300, and approaching that figure in England and elsewhere; inextricably bound up with this economic shift was a movement towards the intensified exploi-tation of the land, greater recourse to the market, and, we may assume, more efficient division of labour.[40] But White failed to make the connexion because the essence of his thesis was that it was the medieval attitude to technological innova-tion that was special and that modern power technology can only have come about through "a medieval mutation in men's attitudes towards the exploitation of nature".[41] He believed he had identified a culturally determined drive to innovate which applied to military hardware, to the tools of agriculture, and to machines of all sorts, but which had nothing to do with any general search for greater efficiency or for new sources of profit.

White succeeded in making medieval technological innovation appear special when in reality it was not. To illustrate that point, let us briefly consider two areas where a progressive development of technique can be observed: construc-tion technology and administration. Neither of these is an activity generally touched upon by historians of medieval technology, but they both provide interesting parallels as well as demonstrating the ability of medieval people to adopt new skills and methods of working when it suited them to do so. There may also be direct points of reference here to those areas of innovation on which historians of technology have tended to concentrate.

To study the surviving examples of medieval architecture and civil engineer-ing is to observe a rising level of expertise from generation to generation, culminating in the increasingly sophisticated and magnificent stone buildings of the later Middle Ages.[42] This is an area where historians of technology might profitably examine the circumstances of the many innovations in technique, whose chronology in the different parts of Europe is well known from the vast numbers of surviving medieval churches. How familiar were the masons with mathematics, and particularly with Euclidean geometry? Or did they constantly revise their received principles of building design by observation, by trial and error, and by word of mouth, independently of any possible scholarly

---

[40] C.C. Dyer, "The hidden trade of the Middle Ages", *Journal of Historical Geography* (1992), **18**: 141–57, on pp. 141–2.

[41] White, *Medieval Technology and Social Change*, p. 79.

[42] R. Morris, *Cathedrals and Abbeys of England and Wales* (London: Dent, 1979), pp. 98–129.

contribution? Was innovation, in this instance, simply a by-product of the masons' increasing ability to put up greater buildings more cheaply, using less building stone? Or was it driven by the constant demands of their ecclesiastical employers for ever grander structures?

A parallel development was the general improvement, through time, of medieval domestic buildings, less spectacular than the advances in church architecture although just as worthy of consideration. There is now abundant archaeological evidence to enable us to describe and date the stages of development from the draughty open halls of the early Middle Ages, made of earth-fast timbers and thatch, to the sound domestic buildings of the fourteenth and fifteenth centuries — built soundly enough, in many cases, to survive until the present day.[43] Constructed with stout timber frames on stone foundations, wattle and daub and plaster walls, fireplaces with chimneys, and often stone or ceramic roof tiles, they could be dry, warm, comfortable, and healthy places in which to live and work, and they were markedly superior to anything that ordinary people had known before in either town or country. Similar developments in the design and construction of agricultural buildings allowed far more effective storage of grain and fodder; the effect must have been to enhance the total benefit from agriculture, so that the soundly built barns of the later Middle Ages also made their contribution to increased productivity along with the more obvious improvments in agricultural technique. An accumulating body of late-medieval archaeological evidence points to capital investment in agriculture at the peasant level, in barns and in corn-drying kilns, presumably with greater involvement in the market in mind.[44]

To take the second example of a well-documented development of technique: with increased secular use of the skills of literacy came new techniques of administration and bureaucracy. The systematic keeping of written records by the English state had begun by 1130; by 1200, crown officials conducted all their essential business in writing, and an effective system for keeping a record of the government's activities was in place. This was not confined to England, but was a general trend throughout western Europe at about that time.[45] The change from an oral to a literate culture had practical consequences: as the civil services of the established states extended their powers and capabilities into evolving centrally controlled systems of taxation and justice, so the power of each state in relation to all its

---

[43] C.C. Dyer, *Standards of Living in the Later Middle Ages. Social Change in England c. 1200–1520* (Cambridge: Cambridge University Press, 1989), pp. 160–9, 200–5.

[44] See for instance, C.C. Dyer, "English peasant buildings in the later Middle Ages", *Medieval Archaeology* (1986), **30**: 19–45; G.G. Astill, "Economic change in later medieval England: an archaeological review", in T.H. Aston *et al.* (eds.), *Social Relations and Ideas. Essays in Honour of R.H. Hilton* (Cambridge: Cambridge University Press, 1983): 217–47, on pp. 232–4.

[45] M.T. Clanchy, *From Memory to Written Record, 1066–1307* (London: Edward Arnold, 1979), pp. 41–59.

subjects, both great and small, was correspondingly strengthened. By comparison with the less ordered societies of early medieval Europe, the states of the later Middle Ages were more stable, providing potentially more favourable conditions for economic enterprise and investment.

At the estate level, too, aristocratic landowners, especially ecclesiastical lords, kept their own records: principally, surveys of their property, but later, and especially in England, detailed records of how the property was managed. The earliest surviving English manorial accounts date from 1208 and simply record the movement of goods and cash, having been devised as a means of ensuring that the demesne officials could not embezzle.[46] But several institutions went on to devise methods of using the account to calculate the annual profit from their estates, for instance the monks of Norwich Cathedral Priory, who began to do so during the 1290s.[47] That these attempts were being made to discover the true profitability of each of the priory's manors points to the active search for savings and greater efficiency that characterized the administration of the greater ecclesiastical estates during the thirteenth century.[48] But at Norwich the exercise was discontinued after 1340, illustrating how the earlier drive for greater production was coming to an end; by the close of the fourteenth century, the almost universal preference of the aristocracy was for letting out their demesne lands on short-term tenancies, so leaving any attempt to increase production to the lessees. And let us remember that this was the very period in which the windmill became such a prominent feature of large areas of the English landscape, before — during the later fourteenth century — it began to decline in numbers. That these windmills were built to satisfy an existing need we have already seen; but just as important a factor in the success of this innovation was the willingness of the improving landlords of the thirteenth century to invest. Equally, the windmill's decline as milling profits fell cannot be dissociated from a change of attitude on the part of the rentier landlords of later centuries. There is, therefore, a clear — if as yet superficial — connexion between the adoption and subse-quent eclipse of a specific technology and what might be called the business ethos of the dominant social class.

# EVOLUTION OF TECHNIQUES BY SMALL PRODUCERS

The crucial question of the extent of innovation from below, from the lower classes of medieval society, was inadequately dealt with in *Medieval Technology and Social Change*. White wrote of "the dynamism of the early medieval peasantry" as a factor in agricultural expansion, but then made no further

---

[46] Clanchy, *From Memory to Written Record*, p. 71.

[47] E. Stone, "Profit and loss accountancy at Norwich Cathedral Priory", *Transactions of the Royal Historical Society* (1962), 4th ser. **12**: 25–48.

[48] See, for example, E. Miller, *The Abbey and Bishopric of Ely* (Cambridge: Cambridge University Press, 1951).

reference to it when he came to discuss changes in land use and agricultural techno-logy. Indeed, he was reluctant to associate any social group with these changes, preferring for the most part to describe them without exploring the identity or motivation of those who might have brought them about.[49] Otherwise, he wrote about technological advances that could be achieved only by those with authority and sufficient resources to expend on new devices and new practices. Yet we know, after John Langdon's research into medieval draught animals, that the important transition from ox power to horse power was a peasant achievement. Blissfully ignorant of the agricultural handbooks, such as Walter of Henley's, which praised the virtues of the ox as an agricultural animal above those of the horse, the peasants were driven by necessity to use the cheapest animals they could get. These turned out to be the elderly demesne horses, prudently sold while they still retained some value; perfectly adequate for the peasants' purposes, they were also more versatile than the ox, and far better suited to the varied needs of the small farmer. Not least, the superior speed of the light horse-cart was to be a factor in giving peasants greatly improved access to the market.[50] This was an example of a major innovation, then, which was achieved without the need for unusual investment on the part of the innovators and which occurred within small-scale enterprise.

In the same way, there is reason to believe that during the fifteenth century a partial withdrawal by the aristocracy from operating mills on many of their man-ors was matched by the construction and operation of horse mills by peasant proprietors. Examples of demesne horse mills come from earlier centuries, when they are also to be found, on occasion, as the property of peasants or of urban bakers and brewers.[51] Without the expensive construction and repair costs of watermills and windmills, horse mills were evidently economical and could be operated profitably as small-scale enterprises. They are less exciting to historians than the more spectacular medieval machines, and consequently have attracted less attention; because they seldom belonged to lords, they were also far more likely to have escaped being recorded. Nevertheless, there is sufficient evidence for their existence to suggest that the peasant horse mill of the fifteenth century was in the long tradition of the horse-powered mills and gins that provided farms, mines, and industrial concerns of all sizes with a convenient source of power until the present century.[52]

There is an obvious need for more sustained research into the technology and methods of peasant agriculture and household enterprise, which are only indi-

---

[49] White, *Medieval Technology and Social Change*, pp. v, 39–78.

[50] J. Langdon, *Horses, Oxen and Technological Innovation. The Use of Draught Animals in English Farming from 1066 to 1500* (Cambridge: Cambridge University Press, 1986); D. Oschinsky (ed.), *Walter of Henley and Other Treatises on Estate Management and Accounting* (Oxford: Clarendon Press, 1971).

[51] Holt, *Mills of Medieval England*, pp. 17–20, 25, 29, 40, 166–7, 169–70.

[52] F. Atkinson, "The horse as a source of rotary power", *Transactions of the Newcomen Society* (1960–1), **33**: 31–55.

rectly recorded but which accounted for the greater part of medieval production. Karl Gunnar Persson has argued that the expansion of the medieval economy is explicable only if we presuppose a steady improvement in basic techniques on the part of the people doing the work: that there was a constant if weak pressure towards technological change, simply as a by-product of practice.[53] As a hypothesis, it is consistent with Langdon's findings, but it requires testing against other large bodies of data. One possibility would be to examine a new industry with a developing technology, for example coal mining: the later Middle Ages saw the emergence — or re-emergence — of a mining industry that laid the foundation for the expanding coal production of the early modern period, and within which miners must have had to evolve new techniques of working.[54] An alternative possibility would be the artefacts mass-produced by medieval craftsmen and found in quantities by archaeologists — pottery, or metal goods such as locks — from which changes in production techniques over time could be detected and measured.

## TECHNOLOGY AND ECONOMIC DEVELOPMENT

If such research is not being done, it is because medieval technology is still neglected by social and economic historians, at the same time as historians of technology have in the main abandoned the Middle Ages to White and his disciples. Many medievalists would still take Sir Michael Postan's dismissive view — formed, it must be stressed, by a lifetime engaged in research into agrarian history — that "the real problem of medieval technology is not why new technological knowledge was not forthcoming, but why the methods, or even the implements, known to medieval men were not employed, or not employed earlier or more widely than they in fact were".[55] Postan was speaking specifically about agricultural technology, but he would surely have extended his judgment to other forms of production as well. While he recognized that the period did see a number of significant advances on both the intellectual and the technical planes, he concluded that, taken all together, this was a poor total by comparison with other periods.[56] And whether or not they would support Postan's position in its entirety, medieval historians would certainly tend to agree that if the people of

[53] K.G. Persson, *Pre-industrial Economic Growth. Social Organization and Technological Progress in Europe* (Oxford: Blackwell, 1988).

[54] J. Hatcher, *The History of the British Coal Industry* (Oxford: Oxford University Press, 1993), vol. 1, pp. 21–30.

[55] M.M. Postan, *The Medieval Economy and Society. An Economic History of Britain in the Middle Ages* (1972; Harmondsworth: Penguin, 1975), pp. 46–7.

[56] M.M. Postan, "Why was science backward in the Middle Ages?", in Postan, *Essays on Medieval Agriculture and General Problems of the Medieval Economy*, (Cambridge: Cambridge University Press, 1973), pp. 81–6.

the Middle Ages did indeed have any genius for innovation, it was in the organization primarily of the human resources rather than the material resources of their society.

Despite all of White's assertions to the contrary, that verdict is well supported by the evidence. The Middle Ages was not a period peculiarly conducive to innovations in technology, nor were such innovations eagerly exploited, except under very special circumstances. The rate of innovation in any society is likely to be related, ultimately, to the rewards the innovator can command; so when production is small-scale and almost entirely for the local market, as it was in the feudal economy, then the material rewards will seldom be very tempting. But that is not to say that technological change had no impact. There is clear evidence of new devices and techniques, whether introduced by the aristocracy, by peasant cultivators, or by craft producers, which must have contributed to the slow process of economic growth. So it is unfortunate that technological change has received scant consideration within the periodic debates among historians concerning the factors in medieval society that ultimately brought about the transition from feudal to capitalist production. In the main, this has been a dis-course either within the circle of Marxist historians, or between Marxists and those taking opposing lines: if anything, the debate has narrowed over the years, so that the Robert Brenner debate of the late 1970s was essentially between those who saw agrarian class struggle as the prime mover in the transformation of feudal society, and those who saw the major demographic crises of the fourteenth and fifteenth centuries as the stimulus to change.[57] Certainly within either of those processes the level of technology — like other material and intellectual resources — would have been a factor, and many of those who responded to Brenner's initial contribution pointed to a lack of innovation in agricultural techniques. But that may be too restricted a view, and more recently there has been at least one suggestion that this debate now requires a reconsideration of the contribution made by developments in technology.[58] Another welcome sign of the broadening of this discourse has been Richard Britnell's examination of the effects of urbanization and commercialization upon medieval economic development and hence upon the eventual transition.[59]

---

[57] T.H. Aston and C.H.E. Philpin (eds.), *The Brenner Debate. Agrarian Class Structure and Economic Development in Pre-industrial Europe* (Cambridge: Cambridge University Press, 1985). See also the *Science and Society* debate of the 1950s, reprinted with additional materials in R.H. Hilton (ed.), *The Transition from Feudalism to Capitalism* (London: Verso, 1978).

[58] C. Harman, "From feudalism to capitalism", *International Socialism* (Winter 1989), **46**, no. 2: 35–87.

[59] R.H. Britnell, *The Commercialization of English Society 1000–1500* (Cambridge: Cambridge University Press, 1993).

Lynn White called his book *Medieval Technology and Social Change*, aiming to demonstrate that the two elements in his title were inseparable, the former being the cause of the latter. He seems not to have considered that the link between them might have worked the other way: that it was actually social changes and pressures that influenced medieval society's choice of technology, something that Marc Bloch recognized in his several articles. Although White expressed his admiration of Bloch, his own preconceptions did not allow him even to acknowledge that Bloch had reached that conclusion. After 1935, Bloch never returned to the subject of medieval technology and innovation, perhaps in part a recognition of the inadequacy, or the intractability, of much of the evidence. It is usually impossible to be certain when a new machine or technique was introduced, or who might have introduced it; with a few exceptions, however important they may be, the transfer of knowledge or of skills is not easily observed by the historian of any period, and certainly not by the medievalist. Nor are the circumstances of the acceptance or rejection of new working practices easily understandable. But these are not sufficient reasons for the long neglect of medieval technology by historians; there is enough documentary and archaeological evidence to allow a new and rational assessment of its relationship to the complex processes of social change, as we have seen. All that is wanting is for historians to move on from the existing and now outdated literature, to bring new life to what ought to be a flourishing field of study.

**Rethinking the Industrial Revolution**

# Law, Espionage, and the Transfer of Technology from Eighteenth Century Britain

*John Harris*

British opposition to foreign attempts to gain new technology which had been developed in this country has been looked at before. Much interest has been centred on the efforts to end the legislation that attempted to prevent the transfer of technology either by the emigration of people or by the export of machinery. A.E. Musson and Maxine Berg have written important chapters or papers in fairly recent times on the machinery issue; David Jeremy has elegantly and adversely assessed the whole programme of transfer prevention while Eric Robinson has looked at the issue through the papers of Boulton and Watt. I shall not venture into the nineteenth century, but look at the question of legislation from its effective beginning in 1719, in contrast to Jeremy, whose study begins in 1780.[1]

---

[1] A.E. Musson, "The 'Manchester School' and exportation of machinery". *Business History* (1972), **14**: 17–50; Maxine Berg, *The Machinery Question and the Making of Political Economy, 1815–1848* (Cambridge: Cambridge University Press, 1980), pp. 203–25; David J. Jeremy, "Damming the flood: British government efforts to check the outflow of technicians and machinery, 1780–1843", *Business History Review* (1977), **51**: 1–34: Eric Robinson, "The international exchange of men and machines 1750–1800", *Business History* (1958), **1**: 3–15. Much of the material used in this article was gathered with the help of grants from the Economic and Social Research Council and the Nuffield Foundation.

## THE LEGISLATION

What was the legislation, and what do we know of its origins and motivation? It began in 1719, and having recently written on the origins,[2] I shall try to condense my findings. There was a wave of anxious applications to government in 1718–19 about two things. One — on which it seems to have been finally decided that the danger was not so serious as the concern expressed — was the potential transfer of English skills abroad by the apprenticeship here of foreign trainees. The other was a remarkable effort to take leading English technology to France; this story is thick with delicious detail, which cannot be recounted here. The famous Scottish financier, John Law, as he rose to a peak of favour under the Orleans Regency and was briefly Controller-General, conceived a plan to take over to France British workmen in a whole series of trades. This he did through Henry Sully, a watchmaker brought up and trained in England, but of French origin and currently living in France. Sully naturally concentrated on enticing clock and watchmakers over, and large numbers went to Versailles and later Saint-Germain. But through collaborators he made it possible for a considerable emigration of woollen workers, metal workers, steelmakers, glassmakers, and shipyard technicians, as well as men from one or two other trades, to take place. Exactly how many men were involved between 1718 and 1720 we cannot be sure, but it would be over 200 and possibly as many as 300. The size of the operation, its extent across trades, and fears that, if unchecked, it would accelerate rapidly, led to petitions from London and several of the industrial districts and to legislation in 1719.

The resulting Act was fundamentally to avert the threat that "many great and profitable branches of ... trades and manufactures may be implanted into foreign countries". It legislated specifically against those enticing workers in wool, iron, steel, brass, and metals, and clock and watchmakers, but also "any other artificers or manufacturers" to go abroad. A first offence attracted a fine of up to £100 and three months imprisonment (continuable until the fine was paid); a second offence attracted a fine to be fixed by the court and a year's imprisonment. The primary target was thus the *enticer*. But workers who had gone abroad to exercise their trades, or teach them, were required to return within six months if requested to do so by British envoys in the country concerned. Refusal would lead to any person ignoring such a request being denied rights of inheritance in Britain, forfeiting all British lands and possessions, and effectively becoming an alien. Where creditable information was laid of efforts to entice workers, or of workers' intentions to emigrate, they could be apprehended. Bail or imprisonment could be required until trial, and then, if they were convicted, security could be demanded against their leaving the country; imprisonment was possible

---

[2] J.R. Harris, "The first British measures against industrial espionage", in I. Blanchard, A. Goodman, and J. Newman (eds.), *Industry and Finance in Early Modern History* (Stuttgart: Franz Steiner Verlag, 1992), pp. 205–25.

till this was provided.[3] The original emigrants under the Law scheme had left England before this legislation, but became liable under the six months' clause. The majority returned, attracted by a £3000 government grant to settle their debts in both countries, though partly driven out by the disgrace of Law and the collapse of his financial support for the colonies of British workers. But a handful of the chief emigrants, having collected a good share of the grant, returned to France.

The decades after 1720 were not marked by large emigrations, although individuals went. Already by the 1730s, some government officials were concerned that there was no law which prevented the export of tools or machines.[4] Legislation on machinery (with one exception)[5] had to wait till 1750. It was then sought by the wool and silk industries on the grounds that there had been recent enticements of woollen and other workers abroad, together with "tools and engines ... or Draughts, Models and Descriptions". Exactly what episodes upset the woollen manufactures of Trowbridge we do not yet know. While in the ensuing 1750 Act cotton, mohair and silk were added to the 1719 list of specific manufactures in which artisans were prohibited from emigrating, cotton was clearly not a main consideration, as it was tools in the woollen and silk industries alone that were singled out in the machinery part of the Act. Anyone exporting "tools and utensils" could be fined £200, and shipmasters knowingly carrying them could be fined £100. But at the same time those seducing workers had the penalty increased to £500 per worker and, for a repetition of the offence, £1000 for each worker.[6]

The Act of 1750 and its origins have many interesting sidelights. The Trowbridge petition had referred to "Engines" as well as tools, and machines are probably intended to be covered by the words "tools and utensils". But we noted that in their petition the Trowbridge men had interestingly also referred to "Draughts, Models and Descriptions of Tools and Engines". These were not eventually included in the Act. We are, of course, well before the burst of machine invention in textiles, mainly in cotton. So what engines were causing anxiety? It hardly seems likely that it was the fly-shuttle, which the West Country did not take up till the 1790s. Kay had taken it to France, where its success was limited; its significant take-up in Lancashire occurred only about 1760. My guess is that attempts to export new types of calenders or presses were at the bottom of the agitation. There was an English finisher established in Copenhagen by 1752, negotiating about a move to France. The calender was critical in giving high finishes to many textiles. It was particularly important at this time because of its recent application to the production of fashionable

---

[3] 5 George I, cap. 27.

[4] P.R.O., S.P. 36, 20.

[5] 7 and 8 William III, cap 20.

[6] 23 George II, cap 13.

"watered" silk, which may explain why wool and silk were allies at this point. In 1752, the French embassy recruited the silk finisher, John Badger, who took the "watering" process to Lyons.[7] Holker, the Lancastrian Jacobite who from the 1750s to the 1780s was, as Inspector General of Foreign Manufactures, the main agent in securing British technology, was himself a calenderer by trade and much involved in their installation in France. English superiority in finishing preceded that conferred by the spinning and weaving inventions.[8]

In 1774 a very similar Act was applied to the export of "Tools and Utensils" for cotton and linen manufacturers, and obviously applied to the new spinning machinery; the jenny had already been imported into France by the Holkers, and the Milnes were approaching John Holker with the Arkwright carding and spinning machinery by the end of the decade. Exporting those items incurred a fine of £200, collecting for export £200, and offending shipmasters paid the same fine.[9] The export of models and plans in woollens, silk, linen, and cotton was consolidated with that of tools and utensils in 1781, and the comprehensive definition was now "machine, engine, tool, press [i.e. calender], paper [possibly meaning the cardboard for pressing] utensil or implement, or any model or plan" of such.[10]

In 1782 blocks, plates, and other equipment for cotton, calico, and linen printing were prohibited from export.[11] This did not derive from anxiety about Britain's Continental rivals. It arose from a scheme blessed by the East India Company to export equipment and workmen to India, where cottons would be printed by the newer British technologies and then imported into Britain. The instigators of the legislation were the calico printers of the London area, who engaged in the importing of Indian cottons in order to print them here. They were soon joined by Lancashire petitioners, however, and the resulting Act prohibited all export of printing equipment to foreign parts, carefully not excluding British possessions.

In 1785 and 1786 there were two Acts relating to the exporting of what were called "iron and steel manufactures", which did not mean the metals themselves or consumer goods made from them, but the equipment of trades making metal goods. It was clearly directed at the machinery and tools of the West Midlands metals industry. Stamps and their mountings, presses of all sorts, piercing machines, dies and die-sinking tools, lathes of all kinds, polishing equipment, silver-plated metals, tools, slitting equipment, and screw-making equipment

---

[7] Archives Nationales, F12 1442.

[8] J.R. Harris, "John Holker: a Lancashire Jacobite in French industry" [first W.H. Chaloner Memorial Lecture], in *Transactions of the Newcomen Society* (1992–3), **64**: 131–42. A useful study is A. Rémond, *John Holker manufacturier et grand fonctionnaire en France au XVIIIᵉ siècle 1719–86* (Paris, 1946).

[9] 14 George III cap 71.

[10] 21 George III cap 37.

[11] 22 George III cap 60.

were main items. Models and plans were again prohibited. It was also specifically said that the increased penalties for suborning under the 1750 Act applied to iron and steel manufactures, and that customs officers successfully bringing prosecutions should share the fines.[12] There was considerable opposition to this Act from iron merchants and manufacturers, led by the great Richard Crawshay, on the grounds that the exclusion of prohibited items would prevent the fulfilling of foreign orders for assorted goods in iron and steel which included any of the prohibited items. Moreover, it was stated, "the said Act betrays a fearful Jealously of Foreigners, which we consider as ill-founded"; failure to export would lead them to manufacture themselves and produce a rapid improvement in their manufacture of iron and steel goods.[13] The result seems to have been that an amending Act was passed in 1786. This Act emphasized that it was lawful to export goods whose export had been permitted before the 1785 Act; to this statement there was added an amended list of items still to be prohibited. Equipment for the casting and boring of cannon was now included, as were wire moulds for paper-making, wheels and other tools for polishing and engraving glass, and the tools of glass-blowers and of saddle- and harness-makers.[14] Annually renewed till 1796, the Act was then made permanent.

## THE LAW: THE CASE FOR FAILURE

Such was the law. Was it in any way effective in the eighteenth century? It is very easy to make a case that it was not. Confining myself solely to France, which I have particularly looked at, one can say that key methods of calendering and bleaching cloths, the fly-shuttle, the spinning jenny, the water-frame and the first examples of the mule, Birmingham small metal trades techniques, the rolling of copper sheathing, the cold rolling of ships' bolts, the first machinery for making ships' blocks, the coke-smelting of iron, and the boring of cast iron cannon from the solid had all gone over by the Revolution, as well as the lead chamber process for sulphuric acid production and flint glass making. All that without looking for trifles. One might presume that, technologically at any rate, the industrial revolution had been packaged and safely delivered in France. There were also the cases where substantial numbers of workers had gone to France. Holker alone had been responsible for groups at Rouen, Sens, and Bourges; there were distinct collections of workers taken for the Birmingham trades by Alcock, De Saudray (two substantial groups), and the Orsel brothers. A "colony", perhaps of two hundred souls, was at the Romilly copper sheathing works in the mid-1780s. There was a steady inflow of professionally expert

---

[12] 25 George III cap 67.

[13] Birmingham Public Libraries, Matthew Boulton Papers, Samuel Garbett Box 1.

[14] 26 George III cap 89.

workers, and makers, of the new textile machinery from the end of the 1770s.[15] So were the laws against the emigration of workers and the export of machinery both idle? Where they were broken, did this mean immediate success for the French?

## THE LAW: THE CASE FOR SIGNIFICANCE, SKILLS, MACHINES, DRAWINGS

First, there could be movements of men and equipment without real success with the technology, or subsequent diffusion. Le Creusot was thought to be the great exemplar of British coal-based technology, but the coke-smelting of iron was a costly flop, a "counter-example",[16] while the architecturally magnificent flint glass works there did not have any notable imitators and was not qualitatively outstanding. The legislation could fail to prevent some artisan emigration and exports of related equipment, but in certain cases what was really needed was a lengthy stay on the part of the original artisan emigrants and a topping up with successors, particularly as the British technologies improved. In the case of the flint glass works, originally sited in Paris, the British embassy officials intervened under the 1719 Act and got most of the English workers to return; only a few others were obtained for the Le Creusot glassworks.[17] At the great ironworks, William Wilkinson was only intermittently present during construction and did not stay to provide long-term management; he seems to have brought few workers with him, and there is no record of their lasting presence. There was, of course, no legislative restriction on businessmen going abroad; there might have been *ad hoc* measures to prevent William Wilkinson going to develop arms industries in war time, but there were none in peace time. As with Le Creusot, the large colony of workers at the great copper works at Romilly melted away, and there were only four foremen left at the end of the century.[18]

While many individuals and groups of workers were suborned to go to France, it is easy to overlook the fact that many returned, for a variety of reasons, such as cultural malaise, mistreatment, the disruptions of the Revolution, or war. The successful seduction of a group of workers did not mean certain or

---

[15] These general flows of technology have been outlined in J.R. Harris, "The transfer of technology between Britain and France and the French Revolution", in Ceri Crossley and Ian Small (eds.), *The French Revolution and British Culture* (Oxford: Oxford University Press, 1989), pp. 156–86, and in "Movements of technology between Britain and Europe in the eighteenth century", in David J. Jeremy (ed.), *International Technology Transfer. Europe, Japan and the USA, 1700–1914* (Aldershot: Elgar, 1991), pp. 9–30.

[16] Denis Woronoff, *L'Industrie sidérurgique en France pendant la Révolution et l'Empire* (Paris: Ecole des Hautes Etudes en Sciences Sociales, 1984), pp. 336–9.

[17] Archives Nationales, F12 1486.

[18] Ibid., F12 1313.

continuous success. Even where the technology concerned was got to work and established at a level that bore some approximation to that then existing in England, it did not advance further without the arrival of additional English experts. (Here, it is worth noting that the failure of the French to add creatively to many technologies imported from England needs more examination.) Hence the failure of the law to prevent some instances of particular techniques reaching France does not mean that it was a complete failure, or that a total absence of legal inhibitions would not have led to better, more complete, and continuously updated transfer.

Let us concentrate for a moment on the emigration of artisans. There is no listing of all the occasions when the law was enforced, still less of those when threats of its enforcement deterred either intending suborners or workers who had been approached to emigrate. There are, however, recorded cases in which suborners were arrested or high bails were demanded, and there are instances of imprisonment: in at least two cases, this led to suicides in English jails. At times, local industrialists issued warnings that they were prepared to enforce the law (a practice that can sometimes be traced in the growing provincial press), and such activities were a common feature of the associations of local industrialists forming in the 1780s. When French manufacturers did their own suborning in England, they sometimes had to get out of the country precipitately when discovered.[19] Coming from an absolutist country, they perhaps failed to appreciate that eighteenth-century Britain was not really a police state and that law enforcement was not always rigorous. But industrial spies like De Saudray and Le Turc, whom we know to have been highly courageous, treated the legislation with great respect, even though they managed to get English workers out of the country. The existence of the laws, too, may have had some influence on the qualities of the men who were prepared to emigrate; those who had small properties, or significant expectations, or who were already well paid would not want to take risks; those who were footloose, or in debt, or discontented and unruly, could be uprooted more easily. Du Pont de Nemours thought English workers could be detached, because, unlike their docile German counterparts, they were "risk takers and greedy".[20] However, once across the Channel their new masters quickly grew exasperated with many "mauvais sujets".

It was understandable that in 1719 the great concentration was on the prevention of emigration and the loss to the foreigner of the skill of key workers. The importance of skill did not decline, but the significance of machinery increased over the century. Although, as is well known, many English machines were taken to France, the acquisition of small numbers of machines of various types

---

[19] For instance, Hyde in 1764 and De Saudray in 1777 both left Birmingham precipitately. See J.R. Harris, *Essays in Industry and Technology in the Eighteenth Century* (Aldershot: Variorum, 1992), pp. 168–9.

[20] Archives Etrangères, Mémoires et Documents Angleterre, 65. Third memoir of Dupont fol. 88.

did not necessarily produce instant results. The view of Dupont De Nemours that new English machines could reach France within six months of their being successfully launched may have been only a slight exaggeration, but that "man of systems" grossly misunderstood the catching up process.[21] The machines had to be matched by men able to erect them and to train French operatives. Nor was it a simple matter to copy machines, particularly as they became more complex, had more metal parts, and ran at greater speeds. In France there were obstacles in the poor quality of iron, the great shortage of steel (there was a total absence of cementation and crucible cast steel production), and a lack both of non-ferrous metals and of experience with alloys. Naturally, also, there was a shortage of French workers with the right skills to build machines, leading to a heavy and lengthy dependence on English machine builders.

These materials and skill shortages led several important engineering witnesses before the Select Committee on Artisans and Machinery of 1824 to state that the French were still well behind the British, in building machines as quickly, as accurately, and as fast-running; and this despite the large number of relevant English workers who went over to France.[22] Napoleonic France had at least 177 Englishmen attached to the textile industry, around sixty of whom are described as "mécanicien".[23]

Although specimen machines could get past the Customs, this was often at the cost of splitting them in parts between different consignments and the consequent trouble and delay of reassembling them. But one machine (even once safely delivered) did not make a factory. The law meant that there was no mass export of machines. If, as a consequence, the French machines that were copied from the samples were costlier, less well made, and slower to operate, the laws against the export of machinery were having some effect. And if the relative superiority of the "mechanic arts", even though improvement had taken place in France by 1824, was still as much in England's favour as it had been fifty years before,[24] then there was some sense at that date in retaining the laws against the export of machinery. It also made sense that the laws should not be relaxed until not only France, but also Belgium and Germany, for example, were sufficiently close in mechanical efficiency for machine building for their own industry not to be a serious problem. By 1840, the boot was on the other foot, and the foolish thing would have been not to supply a growing international market for machinery.

---

[21] Ibid.

[22] Parliamentary Papers. *Report of Minutes of Evidence from the Select Committee on Artisans and Machinery* (1824), First Report, Evidence of Galloway (pp. 18–19) and Donkin (p. 34); Third Report, Evidence of Alexander (p. 102), Booth (p. 160), and Ewart (p. 251).

[23] Margaret Audin, "British hostages in Napoleonic France. The evidence with particular reference to manufacturers and artisans" (M.Soc.Sci. Thesis, University of Birmingham, 1987) and communications from her. Unfortunately Mme Audin did not live to complete her doctoral study on this subject.

[24] See note 22, above.

Some eighteenth-century Frenchmen suggested that if only plans and models or even plans alone could be secured, they would immediately have the means of obtaining British technology. But this facile optimism was rejected by the better and more experienced minds, though they did not dismiss plans and models as useless, seeing them as worthwhile auxiliaries. Hand and eye and the work of someone who had himself practised the techniques were essential. Akos Paulinyi has shown that this was true even for the first half of the nineteenth century, when, as he says, "the overwhelming bulk of British technology was not intelligible through written (or printed) information or through drawings published in Britain" and when "the only way to ... understand new technical solutions (or new combinations of old solutions) was personal contact with new technology".[25]

## ATTITUDES OF INDUSTRIALISTS

Of course, legislation is all the more effective if national opinion is behind it. By the late eighteenth century that opinion was not solid. Among employers there were divisions. Garbett and Wedgwood were steadfastly behind the legislation; Crawshay and John Wilkinson thought that it was not of much use and that it impeded trade, and Wilkinson flouted it; Boulton, for his part, was a perfect weathercock, the wind being self-interest. Others actively supported the legis-lation only when their own technology was threatened. Naturally, for the workmen, the strength and composition of the bribe was critical. But late in the century the keenness of the French government to secure the new spinning inventions may have combined with a sense amongst British workers of elitism, adven-turousness, and participation in a new and unprecedented wave of technological advance. This perhaps made textile machine makers in Lancashire and the adjacent counties more daring and more confident in what they had to sell; they were sure that their talents would still be marketable if they chose to return, and they were, as a result, particularly prone to try their fortunes abroad.

There are anomalies, some of them important. There was no attempt to prevent the export of steam engines, either of the Newcomen or the Watt type, or to stop their erectors from going abroad. Both the Newcomen proprietors and Boulton and Watt took out *privilèges* in France. Where the employment of coal in industry was critical, the French had great difficulty in transferring technologies; furnace building, refractories, crucibles, and even proper stoking eluded them, so that steelmaking — to which they attached the highest importance — was

---

[25] Akos Paulinyi, "Machine tools in the transfer policy of the Prussian 'Gewerbeförderung' (1820s–1840s)", in Dan C. Christensen (ed.), *European Historiography of Technology. Proceedings from the TISC-Conference in Roskilde* (Odense: Odense University Press, 1993), p. 20.

virtually a total failure, with coke-smelting a near disaster.[26] Flint-glass achieved only an imperfect success after a series of ludicrous blunders. The problem here was that the technologies were so alien that there were great problems in knowing whom to recruit for what. The most important hand tool in which the English had an immense superiority was the file, absolutely critical in engineering precision. But its export was not forbidden under the Tool Acts. Why? From the early years of the eighteenth century, files had been exported to France, often smuggled and entering via Holland and Flanders, so that it was probably decided not to interfere with an established trade. Again, the legislation was not, as might be thought, particularly harsh against workers but rather against those seducing them. That it was unjust to prevent a worker from going abroad when a merchant or manufacturer was free to do so seems to have been appreciated; the point was emphasized in the enquiry of 1824 and was a reason for the resulting repeal of the legislation against the emigration of artisans.


## ENGLAND AND FRANCE: THE TECHNOLOGICAL DIFFERENCE

France has been taken as the example of attempted industrial espionage here because Anglo-French technology transfer has been the subject of the writer's work for some years. It may be useful, in conclusion, to give a little general background. The work of François Crouzet, which has had an important impact on other historians, has shown that until the pre-revolutionary crisis, France's economy was developing at a rate at least equal to Britain's in most respects, and this would include much of her industrial output. Admittedly, equivalent speed of development was from a significantly lower base in the early eighteenth century. Nevertheless, when France engaged in technological borrowing — or theft — from Britain, it was not done by a country with a very much poorer economy, but rather by one within the same league. Again, it is only about the end of the seventeenth century that Britain moved from being a technological debtor to the rest of Europe, to being a technological creditor and a target for the acquisition of technologies. Naturally this was not an overnight change, but it was a surprisingly quick and decided one, and the John Law scheme which impelled the answering British legislation against industrial espionage is remarkable testimony to it. In the end, as Crouzet said, "the fundamental difference between the two economies is in the technological field". There is, however, a tendency to presume that this difference commences only

---

[26] J.R. Harris, *Essays in Industry and Technology*, chapter I ("Coal, skills and British industry", pp. 18–24), and chapter III ("Attempts to transfer English steel techniques to France", pp. 78–112).

with the British developments of the late eighteenth century. Some historians have even taken the absurd view that the real significance attaches only to the famous group of textile inventions of the 1760s and 1770s; they even suggest that these are so unprecedented that they could have taken place in any of the more developed European countries and that they gave a quite random and for-tuitous, if decisive, impetus to British industrialization.[27]

Such views miss the importance of the long history of previous technological change in Britain, in which a particularly important theme is the move to the general use of cheaper mineral fuel. This nearly always necessitated important technical change in order to accommodate the use of the equipment of the relevant industry to employ coal and to enable the fuel to be so burnt as to maintain or vary heat at will, which in turn meant that knowledge of coal types, their advantages and problems, had to be gained. The long success with this change of fuel, achieved by many separate advances, large and small, over a couple of centuries, was a major reason for a willingness to try new methods in other industrial fields and to be prised away from traditional practices.

On another front, there is increasingly a justifiable tendency among historians to look for broad cultural reasons to explain English technological fertility and advances. This is to be welcomed, providing that the current attitude to techno-logical change is included in the concept of eighteenth-century culture. Though convinced of the importance of the achievements of the scientific revolution as a valuable and growing element in a technologically favourable culture in Britain, I should yet be reluctant to give this a near-monopoly of cultural influence.[28] Even considering the immense influence of Newton, it would not be easy to claim that British scientists were more important than their French counterparts in the eighteenth century: argument here would have to centre on the extent of the penetration of scientific ideas down the ranks of society, assisted for instance in Britain by subscription lectures and local scientifically oriented societies. There does not appear to have been enough good transnational comparison to make it clear how far France was handicapped here, if at all. Openness to change in general, including technological change, has a certain but unquantifiable relationship to personal freedom of action, thought, and expression. This was

---

[27] François Crouzet, *Britain Ascendant. Comparative Studies in Franco-British Economic History*, trans. M. Thom (Cambridge: Cambridge University Press, 1990), chapters 2 and 3, *passim*, and p. 26. See also N.F.R. Crafts, "Industrial revolution in Britain and France: some thoughts on the question "Why was England first?", *Economic History Review* (1977) 2nd ser. **30**: 429–41. This has been brilliantly and effectively demolished by David Landes, "What room for accident in history?: explaining big changes by small events" [1993 Tawney Memorial Lecture], ibid. (1994), **47**: 637–56.

[28] See for instance Margaret C. Jacob, *The Cultural Meaning of the Scientific Revolution* (Philadelphia: Temple University Press, 1988), for a very perceptive discussion of the place of science in general culture from the seventeenth century and its relation to industrialization.

most emphatically stated by De Tocqueville in a famous passage, in which he related these freedoms indissolubly to British economic success.[29]

When the French (but not only they) set out to purloin new and superior British technology in the face of a body of legislation designed to make their activities illegal, they often discovered marked cultural differences. Those gathering intelligence or directly spying were normally doing so as servants of the French state or as men encouraged and supported by it. They had sometimes to explain to their masters that they were operating across a cultural divide, and that British industrialists, however odd their attitudes seemed, were still serious and substantial people whose technological leadership was indisputable. Such, for instance, was the gist of De La Houlière's report on his successful attempt to recruit the services of the Wilkinson ironmasters.[30]

An important divergence between the technological cultures of England and France can be seen in a difference in the use of language. The wording commonly used in connexion with industrial improvement in mid- and late-eighteenth-century France has, I believe, a noteworthiness which has not been pointed out. It was routine when any inventor, or deviser of a better process, approached the authorities for permit, privilege, subsidy, or other help, for the individual (or those putting his case) to say that his process would bring the particular industry to the point of perfection. There was a stock wording: "give it all that perfection of which it is susceptible". Meeting this cant phrase or jargon again and again, it is very easy to discount it, perhaps even to think of it as repeated to the point of meaninglessness. In Britain, even for major inventions, when the term "perfecting" is used, it is not in this absolute way. I now think this difference in language highly significant. The French had the view that if you pursued industrial improvement by using enlightened criticism of the existing system, abolishing mystery and obscurantism, establishing the proper principles by which the industry should be conducted, and seizing upon the key invention which gave the best method of operation, the technological castle had, so to speak, been finally captured. In Britain, the notion of an achievable finality was absent. By the early decades of the eighteenth century, men had become used, over a great extent of industry, to continuous incremental advance, by large steps

---

[29] Alexis de Tocqueville, *Journeys to England and Ireland*, ed. J.P. Mayer (1958), pp. 115–17. As part of a magnificent purple passage, Tocqueville wrote: "I think it is above all the spirit and habits of liberty which inspire the spirit and habits of trade" (in which he included industry). But he thought this spirit so powerful that it could overcome major resource deficiencies. On this, I differ from him, believing that the richness and complementarity of British resources is seriously neglected by both economic and technologicalhistorians.

[30] W.H. Chaloner, "Marchant de la Houlière's report on casting naval cannon in the year 1775", *Edgar Allen News*, December 1948 and January 1949; reprinted.

and small. In this advance, great victories might be recognized, but they were never complete and always left new technological fields to conquer.

# CONCLUSIONS

My paper has attempted to show that industrial espionage was practised against Britain from the beginning of the eighteenth century, and this has been illustrated by evidence about the main spying nation, France. Laws were framed to combat the espionage from the end of the second decade of the century. At first, the legislation was directed against the seduction of skilled workers abroad; this remained important and was emphasized and extended by later Acts through the remainder of the century. Laws against the export of machines, on the other hand, came in from 1750. These laws were not properly enforced, if only because the means of doing so were limited in a country that was not a police state. While the Act of 1719 in fact prohibited the seduction and emigration of all skilled workers, magistrates tended to act only against artisans and their seducers in trades specifically named in the Act or additionally cited in later Acts. While there was some confusion on the part of both the authorities and workers, skilled men could not be prosecuted for having emigrated unless they had refused to return in six months after notice was given them. The Tools Acts did not, as has sometimes been thought, include the tools, hand or machine, of all trades, but principally those of the Birmingham hardware district; some of those listed in the 1785 Act were removed in that of 1786, and others relating to some other trades added.

Yet, when all these limitations and exceptions are taken into consideration, there is still good reason to believe that the relevant laws had some deterrent effect, in inhibiting, and sometimes completely disrupting, foreign attempts to secure British technologies and British workers who would practise and teach them abroad. The laws against the export of machines and equipment could hardly prevent specimen machines from going abroad, but that did not mean that the full provision of foreign manufactures by the mass export of British machines was possible. There were well-established differences in technological culture between England and France before the middle of the eighteenth century; to some extent, this reduced the effectiveness of the espionage process by a reduced understanding of the nature of the technology to be transferred, and the precise circumstances in which it might be successful once transferred. The existence of the laws against industrial espionage meant that investigation into what was happening in British industry had to be hurried, secretive, even fearful on occasions, not the best conditions in which to fully absorb a novel technology. It was not easy for European observers to understand a British technological culture which, partly because of the conquest of the problems of a demanding new industrial fuel, had absorbed the notion of continuous change rather than that of perfectibility; a lack of comprehension here affected, in turn, the mental horizon of those trying to steal British technology by espionage. It may be tentatively suggested — to return to my comment earlier in this paper — that this had something to do with the failure of the French to make significant improvements to the British technology which they did acquire.[31]

There was no rigid uniformity of opinion in Britain. As we have seen, some industrialists in the 1780s pressed hard for the introduction and implementation of laws against machine export, and others resisted them. The penalties against those seducing workers were extremely severe, and there are indications that their very severity made magistrates and juries hesitant about their imposition; penalties for workers seduced, by contrast, were only a matter of finding sureties for their remaining in the country. By the end of the French wars, the injustice of preventing workers from bettering their lot abroad was widely felt and it was righted ten years later. The machinery export issue then became increasingly a quarrel between those manufacturing textiles by machine who hoped that depriving foreigners of the best British built machines would impose a technological lag on them, and machine builders keen to enter a booming export market while they still possessed a technological premium. The rather specialized arguments of the time about the wisdom or effectiveness of damming the flood should not cause us to lose sight of the men and measures of the previous century, when Britons tried to safeguard their new experience of technological leadership.

---

[31] For instance, the steam engine. "A partir de 1789, la construction des machines à vapeur n'a plus évolué en France. Les machines construites à la fin de l'Empire seront exactement celles que Watt fabriquait en 1785"; see Jacques Payen, *Capital et machine à vapeur au XVIII^e siècle. Les frères Périer et l'introduction en France de la machine à vapeur de Watt* (Paris and The Hague: Mouton, 1969), p. 176.

# Concepts of Invention and the Patent Controversy in Victorian Britain

## Christine MacLeod

In 1827 Samuel Crompton, inventor of the spinning mule, died impoverished and was buried in an unmarked grave. In 1833 Richard Trevithick, pioneer of high-pressure steam and the locomotive, was spared the indignity of a pauper's funeral only by the generosity of his fellow workmen. Yet in 1859 Crompton was not only celebrated in a full-length biography, but his biographer, Gilbert French, also raised a subscription of £200 for a monument over his grave in Bolton churchyard and a further £2,000 for a copper-bronze statue, formally presented to Bolton town council, with much pomp and circumstance, in 1862. And in 1883, by public subscription, a Trevithick memorial window was installed in the north aisle of Westminster Abbey, and a Trevithick engineering scholarship was endowed at Owens College, Manchester.[1] The heroic inventor penetrated yet further into the cloisters of academe. Christ's College, Cambridge, when glazing its new dining hall in 1879, chose to celebrate William Lee, the late-sixteenth-century inventor of the stocking-knitting frame. His precise academic pedigree is still disputed between Oxford and Cambridge, and it is not certain that he attended either.[2] Nonetheless, Christ's claimed Lee as its

---

[1] Leslie Stephen (ed.), *Dictionary of National Biography* (London: Smith, Elder & Co., 1885–1903), "Samuel Crompton (1753–1827)", "Richard Trevithick (1771–1833)"; Gilbert French, *The Life and Times of Samuel Crompton, Inventor of the Spinning Machine Called the Mule* (London and Manchester, 1859); Michael E. Rose, "Samuel Crompton (1753–1827), inventor of the spinning mule: a reconsideration", *Transactions of the Lancashire and Cheshire Antiquarian Society* (1965), **75**: 11–32.

[2] Science Museum Library, London [SML], Bennet Woodcroft MSS, Z27/B, ff. 227, 229; and see below, note 5.

own, so there he stands, holding a model of the knitting frame, amidst the statesmen and assorted intellectuals who fill the other panes. These are probably the most spectacular instances of a phenomenon that marks out the second half of the nineteenth century as an exceptional period when inventors, most of them long dead and neglected in their lifetimes, were elevated to heroic status.

Invention had remained a largely anonymous enterprise before the mid-nineteenth century. The merits of rival claimants to priority of invention had been disputed in industrial histories, such as Baines's *History of the Cotton Manufacture in Great Britain* (1835) and Ure's *The Cotton Manufacture of Great Britain* (1836), which pursued the often vituperative debate over Richard Arkwright's reputation.[3] Arkwright and James Brindley were deemed worthy of a chapter each in a Smilesean anticipation, Richard Davenport's *Lives of Individuals Who Raised Themselves from Poverty to Eminence or Fortune* (a title with rather less panache than *Self Help*), but little was made of their respective inventions; both, as heroes "in an existential struggle with nature", attracted the brief notice of Thomas Carlyle in the early 1840s.[4] The tale of William Lee's invention had entered the folklore of the East Midlands hosiery districts and was repeated, with variations of emphasis and embroidery, by local historians during the eighteenth and nineteenth centuries; in 1847 he became the subject of a romantic genre painting.[5]

With the exceptions of James Watt and Edward Jenner, however, no British inventor or engineer had been the individual subject of a monograph before the

---

[3] Edward Baines, Jr., *History of the Cotton Manufacture in Great Britain* (London, 1835); Andrew Ure, *The Cotton Manufacture of Great Britain* (2 vols., London, 1836). For a full account of this debate, see David J. Jeremy, "British and American entrepreneurial values in the early nineteenth century: a parting of the ways?", in R.A. Burchell (ed.), *The End of Anglo-America. Historical Essays in the Study of Cultural Divergence* (Manchester and New York: Manchester University Press, 1991), pp. 34–9. For similar disputes over James Watt, see Hugh Torrens, "Jonathan Hornblower (1753–1815) and the steam engine: a historiographic analysis", in Denis Smith (ed.), *Perceptions of Great Engineers. Fact and Fantasy* (London: The Science Museum, 1994), pp. 23–34.

[4] Richard Alfred Davenport, *Lives of Individuals Who Raised Themselves from Poverty to Eminence or Fortune* (London: Thomas Tegg, 1841); Carlyle is quoted in Simon Dentith, "Samuel Smiles and the nineteenth-century novel", in Smith (ed.), *Perceptions of Great Engineers*, pp. 52–3.

[5] *Felkin's History of the Machine-wrought Hosiery and Lace Manufactures*, ed. S.D. Chapman (Newton Abbot: David & Charles, 1967), pp. 26–38, 51; James H. Quilter and John Chamberlain, *Frame-work Knitting and Hosiery Manufacture* (3 vols., Leicester: "Hosiery Trade Journal", 1911), vol. 1, p. 2.

1850s.[6] In 1845 Crompton's grandson wrote to Lord Brougham to request that he include his grandfather in the new biographical dictionary being published under Brougham's auspices. A similar request to the editor of Rose's *Biographical Dictionary*, said Crompton, had been ignored, and literary men in general were unwilling "to consider particulars of the life of a man who left no writings behind him, who was a member of no scientific society and (which is the greatest failing in the eyes of Englishmen) who bequeathed no wealth to his family". Even eminent writers, he complained, were astonishingly ignorant about inventors: Alison, in his history of the French Revolution, had attributed the mule to Arkwright and the jenny to Cartwright.[7] The situation was well summarized by a reviewer of Smiles's *Industrial Biography* in 1863: "Mr Smiles rescues no name, but many histories, from oblivion. His heroes are known and gratefully remembered for the benefits they have conferred on mankind, but our knowledge of our benefactors has hitherto been mostly confined to our knowledge of the benefit".[8]

Crompton's complaints were soon addressed by the outpouring of hagiographic lives of inventors, commemorative plaques, and statues during the next few decades. Yet the heyday of the inventor was short-lived. He quickly returned to anonymity. By the time of Trevithick's exaltation in Westminster Abbey, in 1883, interest was beginning to wane. Successful inventors, engineers, and scientists of the later nineteenth century, such as Sir Charles Parsons, William Siemens, and Lord Kelvin, were recognized in their own time with honours from the state, the universities, and professional bodies, and, shortly thereafter, by that ultimate accolade — a place in the *Dictionary of National Biography*. But their posthumous popular fame is slight. It seems that everyone knows the name of

---

[6] Watt had been celebrated in D.F.J. Arago, *Eloge historique de James Watt* (Paris, 1834), which was translated into English twice in 1839. For early biographical notices of Watt, see Torrens, "Jonathan Hornblower". Jenner's biography, by John Baron, was written at the request of his family: W.R. Lefanu, *A Bio-bibliography of Edward Jenner, 1749–1823* (London: Harvey & Blythe, 1951), p. 149; for earlier short notices and later biographies, see ibid., pp. 143–50. Thomas Telford had been persuaded to write an autobiography, published posthumously as *The Life of Thomas Telford, Civil Engineer, written by Himself*, ed. John Rickman (London, 1838).

[7] University College, London [UCL], Brougham MSS 32,998. The references are to *A New General Biographical Dictionary, Projected and Partly Arranged by Hugh James Rose, B.D.* (12 vols., London, 1848); Sir Archibald Alison, *History of Europe during the French Revolution* (10 vols., Edinburgh: W. Blackwood; London: T. Cadell, 1833–42).

[8] *Edinburgh Daily Review*, reprinted in an advertisement for *Industrial Biography. Iron Workers and Tool Makers* (London, 1863), in the endpapers of Samuel Smiles, *Lives of the Engineers. The Locomotive. George and Robert Stephenson* (London: John Murray, 1877). Smiles himself, reviewing a new biography of Watt in 1858, commented on the dearth of biographies of "distinguished inventors": see *Quarterly Review* (1858), **104**: 411–51, on p. 411.

the inventor of the separate condenser (even though most are vague about exactly *what* Watt invented); few, however, know the name of the steam turbine's inventor. The heroic account of invention had been securely established and earlier cohorts of inventors rescued (selectively) from obscurity; ironically, Parsons's generation was ignored by the popular biographers.

This neglect of the great figures of late-nineteenth and twentieth-century technology is not, however, my prime concern in this paper. Rather, my focus is on the third quarter of the nineteenth century, which, through its feverish interest, set the agenda for much subsequent history of technology in this country. Not only did it enshrine the pantheon of subjects for repeated biographical studies,[9] but it also established the biographical, heroic approach itself, which survives to this day in the popular literature and many school textbooks, although long demoted from the academic historiography.


# THE PATENT CONTROVERSY

This heroic approach has either been taken for granted, or assumed to be the singlehanded creation of Samuel Smiles. I shall argue, however, that the brief apotheosis of the inventor was largely the product of a fiercely fought defence of the patent system against a tenacious and articulate campaign for its abolition. Though primarily a battle between vested interests, couched in overwhelmingly pragmatic terms, the protagonists in this "patent controversy" drew on competing explanations of invention to lend some theoretical support to their respective campaigns.

By its very nature, the patent system rewarded *individual* achievement. Its supporters and beneficiaries had to profess the uniqueness of that achievement: they had to argue that without the creative abilities of a particular inventor the invention would never have been made, or, at the least, would have been long delayed. Therefore the inventor merited reward. Its detractors were obliged to play down the role of the individual inventor. They argued that social and economic needs prompted the invention of technological solutions, that invention was incremental, and that the frequent simultaneity of inventions demonstrated their case.[10] While today our intellectual sympathies may lie largely with the determinist arguments of the abolitionists, it is important to recognize that, then as now, their opponents enjoyed significant advantages in canvassing public support. Theirs potentially were the colourful stories of

---

[9] R. Angus Buchanan, *The Engineers. A History of the Engineering Profession in Britain, 1780–1914* (London: Jessica Kingsley Publishers, 1989), pp. 16–19.

[10] For a compendium of abolitionist arguments, see [Robert McFie (ed.)], *Recent Discussions on the Abolition of Patents for Inventions in the United Kingdom, France, Germany, and the Netherlands* (London: Longmans, Green, Reader, and Dyer, 1869).

inspiration and of obstacles overcome by force of character. At their disposal were not only the Romantic cult of genius and heroism, but also the middle-class, commercial ethos that sought to establish the arts of peace above the atavism of war and conquest.[11] The time was ripe for a new set of heroes. In contrast, the abolitionists' case was likely to seem dry, unexciting — and even politically radical.

If the best tunes all belonged to one side, it was the patent controversy that prompted their composition. Before the 1850s, with one or two exceptions, inventors had not been recognized as popular heroes: invention had remained a largely anonymous activity. Moreover, theoretical debate concerning the nature of invention had been, to say the least, muted. It rarely arose outside the jousting of patent litigation, where both litigants and lawyers had to assume that invention deserved individual reward; the courts' task consisted only in adjudicating between rival claimants. However, the individualistic account of invention expounded in the heat of courtroom debate was slow to provide the under-pinning of a full-blown heroic explanation. While the patent system implicitly endorsed and promoted an ideology of individualism, it was the threat to the system's existence in the third quarter of the nineteenth century that called forth a more explicit and much stronger account.[12] This heroic ideology elevated the successful inventor to a creative genius, insisting on a place for him in the national pantheon which had previously acclaimed only statesmen, admirals, generals, and major literary and intellectual figures — with, more recently, Edward Jenner and James Watt, being the exceptions.

I shall briefly describe the "patent controversy", before proceeding to outline the rival explanations of invention advanced by the two camps. I shall argue that the success of the campaign to defend the patent system, enhanced by the popularization of heroic inventors, has obscured from view a resilient discourse of determinism. This deterministic account of discovery and invention was prevalent until the eighteenth century but thereafter largely lost ground to more individualistic versions until "rediscovered" and promoted in this century by Ogborn and Thomas, Gilfillan, and other members of the Chicago School.[13] Though submerged, it was not defunct, and its challenge also provoked a more thoughtful response from some proponents of the "heroic" case.

---

[11] Stefan Collini, Donald Winch, and John Burrow (eds.), *That Noble Science of Politics. A Study in Nineteenth Century Intellectual History* (Cambridge: Cambridge University Press, 1983), pp. 27–8; A.O.J. Cockshut, *Truth to Life. The Art of Biography in the Nineteenth Century* (London: Collins, 1974), p. 122; Arnold Thackray, "Natural knowledge in cultural context: the Manchester model", *American Historical Review* (1974), **79**: 672–709.

[12] Discussion of inventiveness was absent from debates surrounding the reform of the patent system in the 1820s and 1830s: see, for example, *Hansard*, XV (1826), cols. 70–6; XXI (1829), cols. 598–608.

[13] See the bibliography in S. Colum Gilfillan, *Inventing the Ship* (Chicago: Follett Publishing Co., 1935), pp. 165–75.

First, to consider the patent controversy, distilling here the lucid account by Moureen Coulter in her recent book, *Property in Ideas*.[14] The patent system that had emerged since the seventeenth century was reformed by the Patent Amendment Act of 1852. This simplified and cheapened the procedure for obtaining a patent, satisfying many of the objections voiced by patentees and their supporters during the previous half century. However, it also stirred up a hornet's nest of opposition that called into question the system's very existence during the next thirty years. The issues were debated in both the national and the technical press and through a host of institutions, from local Chambers of Commerce and Mechanics' Institutes to national bodies, such as the Royal Society of Arts and the Institutions of Mechanical and Civil Engineers. Coulter has counted at least eighteen monographs defending the patent system published between 1862 and 1877. The abolitionists secured a Royal Commission in 1862 and agitated their way through another Select Committee and a host of parliamentary debates, but by the late 1870s, their star was waning.[15] They had been unable to win over any national body to their cause, prompting such bodies instead to lobby for further reforms to the system, many of which were enshrined in the 1883 Act, which hammered the final, patent nail into the abolitionists' coffin.

The parliamentary select committee which investigated the workings of the patent system in 1851 had heard thirty-three witnesses, of whom only eight recommended its abolition. The impact of the abolitionists was, however, much greater than their numbers alone suggest. First, they included such major figures as Isambard Kingdom Brunel and William Armstrong, both highly respected and inventive engineers, and William Cubitt, the President of the Institution of Civil Engineers. Secondly, Lord Granville, the select committee's chairman, when presenting its recommendations for the Patent Amendment Act in 1851, startled parliament by confessing that he had been convinced by the abolitionists' arguments: "the whole system", he announced, "was unadvisable to the public, disadvantageous to inventors, and wrong in principle".[16] Thirdly, the 1852 Act spawned a vociferous vested interest for abolition among the sugar manu-facturers. By excluding the colonies from the purview of British patents, the Act threatened sugar refiners with the payment of royalties from which their Caribbean competitors would be exempted. Abolition became virtually a

---

[14] Moureen Coulter, *Property in Ideas. The Patent Question in Mid-Victorian Britain* (Kirksville, Mo.: Thomas Jefferson Press, 1992). See also F. Machlop and E. Penrose, "The patent controversy in the nineteenth century", *Journal of Economic History* (1950), **10**: 1–29; Victor M. Batzel, "Legal monopoly in Liberal England: the patent controversy in the mid-nineteenth century", *Business History* (1980), **22**: 189–202.

[15] *Royal Commission to Inquire into the Working of the Law Relating to Letters Patent for Inventions*, British Parliamentary Papers, [hereafter P.P.] 1864, XXIX; *Select Committee on the Law and Practice of Grants of Letters Patent for Inventions*, P.P., 1871, X, and 1872, XI.

[16] *Hansard*, 3rd ser. CXVIII (1851), col. 16.

personal crusade for Robert McFie, the Liverpool sugar refiner, to the point of his entering national politics in its pursuit; elected in 1868, he lost no opportunity to argue the abolitionists' case for the next six years.[17]

However, there were many more who had a vested interest in the retention of the patent system, not least thousands of patentees (their number trebling in the wake of the 1852 Act) and the many manufacturers who had bought, and were operating, patented inventions. There was also a small but articulate body of professional patent agents and barristers specializing in patent litigation.[18] And, at the head of the newly formed Patent Office was Bennet Woodcroft, a man whose role in championing the cause of inventors and patentees was, it appears, pivotal. Woodcroft was tireless in campaigning for a better deal for patentees, maintaining the pressure after 1852 both for still cheaper patents and for better facilities at the Patent Office: the discomforts of the search room were notorious, meriting its nickname, "the drain pipe".[19] Another, most influential champion of the patent system was Henry, Lord Brougham, the prominent Whig politician, advocate of educational and legal reform, and first president from 1857 of the Social Science Association, a man without any vested interest, but with a long history of campaigning for reform of the system to benefit patentees. Brougham's concerns paralleled those of Woodcroft; he had been closely involved in securing posthumous, national recognition for James Watt.[20]

Woodcroft, however, strove to raise the status of inventors as a whole and to revive the memory of many who were dead and forgotten. He combed the country for portraits to establish a projected National Portrait Gallery of Inventors and for machinery and models with which to furnish the Museum of Patents (inconveniently situated at South Kensington, a long walk from the Patent Office in Chancery Lane). He encouraged living inventors to have their portraits painted for the collection and to have their biographies written.[21] With help from Woodcroft, John Timbs produced in 1860 the first composite biography of inventors, "from Archimedes to Isambard Kingdom Brunel" (its emphasis, however, being overwhelmingly modern and British!), which praised the achievements of individual "genius". Its general tenor may be deduced from a passage in the preface: "In tracing the fortunes of inventors and discoverers, it is painful to note how many have become 'martyrs of science'; a phrase ... which, there is reason to hope, will at no very distant time be inapplicable. A

---

[17] *Dictionary of National Biography*, 3rd Supplement, "Robert Andrew McFie (1811–1893)".

[18] H.I. Dutton, *The Patent System and Inventive Activity During the Industrial Revolution, 1750–1852* (Manchester: Manchester University Press, 1984), pp. 34–56, 86–100.

[19] Ibid., frontispiece.

[20] UCL, Brougham MSS, correspondence with James Watt, Jr., and with numerous inventors and patentees concerning reform of the patent system or seeking his assistance.

[21] SML, Bennet Woodcroft MSS, Z27B, ff. 329, 511.

brighter era is at hand".[22] Timbs commended the recent honouring of Jenner by a statue in Trafalgar Square, London, and of George Stephenson in Liverpool and London, at the same time as he deplored the delays in such tributes occasioned by earlier national ingratitude.[23] Woodcroft also hoped to stimulate both civic authorities and individuals throughout the country to rediscover and celebrate local inventors. In 1863, for example, he tried to shame the Lancashire town of Bury into celebrating John Kay, inventor of the flying shuttle, by pointing to Bolton's recent commemoration of Samuel Crompton, and to shame Britain as a whole by reference to France's elaborate honouring of Jacquard.[24]

Ideally, one would quote explicit evidence from Woodcroft's correspondence of his having adopted a conscious policy to help defend the patent system by lionizing inventors. If he ever did consciously adopt such a stratagem, the correspondence is silent, though it should be remarked that two potentially pertinent volumes relating to the 1850s are missing from the Science Museum Library's collection. The evidence admittedly remains circumstantial, and alternative explanations are conceivable for the mid-century's unprecedented, yet brief, flowering of interest in inventors. The euphoria generated by the Great Exhibition and the personal interest of Prince Albert in matters scientific and technical would obviously be explanatory contenders. We might also remark the rising interest in biography as a literary and historical genre, and in particular "the scholarly historical interest in the heroes of the scientific revolution that began to emerge in Britain during the early decades of the nineteenth century".[25]

Whether or not he consciously intended a public relations campaign in defence of the patent system, Woodcroft's efforts certainly promoted the cause. For nearly thirty years he corresponded with an expanding network of patrons, artists, engineers, and biographers (including Samuel Smiles),[26] stoking their enthusiasm to celebrate the virtues of a forgotten generation of inventors; this at a time when his newly created Patent Office was under threat of abolition and desperately in need of better accommodation. He may have tapped a sense of guilt, lurking beneath the triumphalism of the Great Exhibition, that the

---

[22] John Timbs, *Stories of Inventors and Discoverers in Science and the Useful Arts* (London, 1860), pp. v, xii; SML, Bennet Woodcroft MSS, Z27B, f. 246.

[23] Timbs, *Stories*, pp. 130–1, 299.

[24] Bennet Woodcroft, F.R.S., *Brief Biographies of Inventors of Machines for the Manufacture of Textile Fabrics* (London, 1863), pp. 6, 29–31.

[25] Richard Yeo, "Genius, method, and morality: images of Newton in Britain, 1760–1860", *Science in Context* (1988), **2**: 257–84 (258, 259–60, 265–7). See also Cockshut, *Truth to Life*, p. 11; A.N.L. Munby, *The History and Bibliography of Science in England. The First Phase, 1837–1845* (Berkeley, CA: University of California Press, 1968).

[26] SML, Bennet Woodcroft MSS, Z27A, f.197; *Select Committee on the Patent Office Library and Museum*, P.P., 1864, XII, p. 19.

progenitors of Britain's wealth and power had been shabbily treated. "What did the public do for inventors?", declaimed Dr Walkley, M.P., in 1851. "They often treated them with scandalous neglect. What was done for Harvey or for Jenner, of his own profession? Jenner's family were absolutely in want; there were no monuments to his memory, and his name was almost unknown in this country".[27] This was a singularly poor example to choose, Jenner being better rewarded and commemorated than most, but the general point was sound.

Neither side in the debate was able to draw on an extensive or coherent body of thought concerning the nature of invention. There were a few relevant pamphlets, some short passages in books concerned with broader issues, and arguments advanced in the course of patent litigation, but there had been no previous public debates in which each side pressed the other to clarify and defend its premises.[28] It is evident when reading some late-eighteenth and early-nineteenth-century accounts of invention that their authors had not been required to give serious consideration to the role of the individual inventor. Unselfconsciously, they might hint at both individualistic and deterministic ideologies. Gravenor Henson, for example, in 1831 could credit various inventors with "genius" or "ingenuity", and yet allow that "Nothing is more difficult than to decide with certainty, who is the real inventor of machinery, as, perhaps, there are several persons, at the same time, scheming and trying experiments, to accomplish a given end".[29] Even Arago, in his great encomium on the inventive "genius" of James Watt, gave an occasional role to "the mere spur of necessity; for 'necessity is the mother of invention'".[30] This ambiguity was a luxury not permitted to the generation after 1851, which appreciated the significance of this issue for the future of the patent system.

## THE DETERMINIST IDEOLOGY

The determinist argument advanced by the abolitionists will have a familiar ring. It emphasized the collaborative and incremental nature of invention. It claimed that those men generally thought of as major inventors were nothing more than the last

---

[27] *Hansard*, 3rd ser. CXVIII (1851), col. 1548.

[28] Compare this with the fierce, contemporary debate in the historiography of science; see Yeo, "Genius, method, and morality", pp. 261, 267–78.

[29] Gravenor Henson, *The Civil, Political and Mechanical History of the Framework Knitters, in Europe and America* (Nottingham, 1831), ed. S.D. Chapman (Newton Abbot: David & Charles, 1970), pp. 276, 304, 331, 342.

[30] D.F.J. Arago, *Life of James Watt* [trans. Hyde Clark] (3rd edn., Edinburgh: A. & C. Black, 1839), p. 55.

links in the causal chain, putting the final piece into the jigsaw which others had almost completed. And if the particular individual had not been there to insert the last piece, somebody else would have been — perhaps not always at the same time, but before long. Moreover, once the final stage had been reached, there were usually many individuals seeking out the last piece of the puzzle, and so the race to the finish was often a close one.[31] In support of this argument, its proponents cited simultaneous inventions, as Ogborn and Thomas were to do in the 1920s,[32] though rarely did they offer any specific examples. Earl Granville told the House of Lords in 1851 that "the concurrence of similar inventions is very remarkable". Lord Stanley, who had chaired the Royal Commission of 1862–4, also believed by 1869 that the patent system did "more harm than good". Its essential flaw, he thought, lay in the difficulty of rewarding the right person, for "it might happen, and often did happen, that two or three, or even half-a-dozen men, quite independently of one another, would hit upon the same invention".[33]

It followed that a sociological, rather than an individualistic, explanation was required: it was usually framed in terms of social "needs". Though he may be suspected of inventing the proverb, the barrister John Coryton encapsulated this view in 1874 with the claim "it is an old saying, that 'it is society that invents'".[34] One member of his audience, Mr Anderson Rose, was moved to respond that "There was ... no such thing as invention; if the world wanted anything it came to the front in two or three places at once".[35] The evidence of Sir William Armstrong, past president of the Institution of Mechanical Engineers, to the Royal Commission in 1863 made the whole business seem ridiculously easy: "As soon as a demand arises for any machine or implement or process, the means of satisfying that demand present themselves to very many persons at the same time". Warming to his subject, Armstrong continued: "the great majority of inventions are the result of mere accident — if you let them alone they will turn up of themselves".[36] It followed that the fourteen-year

---

[31] *Hansard*, 3rd ser. CXVIII (1851), col. 14; CXCVI, col. 895.

[32] W.F. Ogborn and D. Thomas, "Are inventions inevitable? A note on social evolution", *Political Science Quarterly* (1922), **37**: 83–98.

[33] *Hansard*, 3rd ser. CXVIII (1851), col. 16; CXCVI (1869), cols. 904–5. See also J.E. Thorold Rogers, "On the rationale and working of the patent laws", *Journal of the Statistical Society of London* (1863), **26**: 121–42, on p. 125.

[34] John Coryton, "The policy of granting letters patent for invention, with observations on the working of the English law", *Sessional Proceedings of the National Association for the Promotion of Social Science* (1873–4), **7**: 163–90, on p. 168.

[35] Ibid., pp. 186–7.

[36] *Report of the Commissioners appointed to Inquire into the Working of the Law Relating to Letters Patent for Inventions*, P.P., 1864, XXIX, pp. 414, 415. See also *Hansard*, 3rd ser. CXCVI (1869), col. 895.

patent monopoly was much too great and too exclusive a prize to award the one who only happened to breast the finishing tape first — and in such an easy race! Probably the most cogent explanation of simultaneity came from Isambard Kingdom Brunel, reflecting in 1851 on his own experience; he emphasized, above demand, the cumulative supply of knowledge available in the public realm and the systemic nature of technology which directed inventors' efforts to what Thomas Hughes has termed "critical problems".[37] Brunel commented:

> I believe that the most useful and novel inventions and improvements of the present day are mere progressive steps in a highly wrought and highly advanced system, suggested by, and dependent on, other previous steps, their whole value and the means of their application probably dependent on the success of some or many other inventions, some old, some new. I think also that really good improvements are not the result of inspiration; they are not, strictly speaking, inventions, but more or less the results of an observing mind, brought to bear upon circumstances as they arise, with an intimate knowledge of what already has been done, or what might now be done, by means of the present improved state of things, and that in most cases they result from a demand which circumstances happen to create.[38]

The abolitionist argument had to deny a natural right to property in invention, and therefore to refute the associated point that a patent of invention was analogous to copyright in a book or work of art. It was a case that followed naturally from the simultaneity argument: books and works of art were clearly unique, creative acts, whereas if many minds could reach the "right" solution to a technical conundrum there could be no claim to uniqueness nor any entitlement to property in the solution. "Copyright applied to a creation", Sir Roundell Palmer (another "convert" since 1851) told the Commons in 1869. "But in the case of invention and discoveries, the facts with which they were concerned lay in nature itself, and all mankind who were engaged in pursuits which gave them an interest in the investigation for practical purposes of the laws of nature had an equal right of access to the knowledge and the practical applications of those laws".[39]

---

[37] Thomas Hughes, "The dynamics of technological change: salients, critical problems, and industrial revolutions", in Giovanni Dosi, Renato Giannetti, and Pier Angelo Toninelli (eds.), *Technology and Enterprise in a Historical Perspective* (Oxford: Clarendon Press, 1992), pp. 97–118.

[38] Isambard Brunel, *The Life of Isambard Kingdom Brunel, Civil Engineer* (London: Longman & Co., 1870), p. 492. See also J. Stirling, "Patent right", in [McFie], *Recent Discussions*, p. 119.

[39] *Hansard*, 3rd ser. CXCVI (1869), col. 898.

From this there developed a common abolitionist argument which conflated scientific and technological achievement in a platonistic conception: laws of nature and new technologies were both waiting "out there" to be unearthed; persistent investigators would discover a pre-existent truth or the ideal form of a technical solution. And where the two were not conflated, invention tended to be downgraded to that mere "application of science" which is all too familiar. In neither case could it be right that the first discoverer or adopter should appropriate what belonged to all. John Coryton told his (largely unsympathetic) audience at the National Association for the Promotion of Social Science in 1874 that "Discoveries of great principles are almost confined to men of scientific occupations. Such men — patient and unselfish workers, of whom Faraday may be taken as a type — are ... [the] true improvers of our manufactures. It is to their researches, in the vast majority of cases, that the patentee is indebted for the invention of which he has obtained the exclusive use".[40] In Coryton's view, the selfless, collaborative scientist strives and discovers; the grasping patentee, skimming off the cream, applies and appropriates. The stereotypes live on.

What was the source of these deterministic views? They represent the secular legacy of the providentialist account of invention which had held sway until the eighteenth century.[41] Renaissance neo-platonists suggested there was a stock of inventions to be released and materialized at appropriate times, provided humankind made an effort to discover them — implying both need and desert. The word "invention" was used in a way closer to its classical root, in the sense of uncovering something which had been there all the time, on a par with new lands or planets. In his Boyle lecture of 1712, William Derham depicted the inventor as subject "to the agency or influence of the spirit of God", acting as God's agent "to employ the several creatures; to make use of the various materials; to manage the grand business". Some things, flight for example, remained unachieved, said Derham, "because the infinitely wise Creator and Ruler of the World has been pleased to lock up these things from man's understanding and invention, for some reason best known to himself, or because they might be of ill consequence, and dangerous amongst men".[42] On the other hand, as many Protestants argued, the timely invention of printing had

---

[40] Coryton, "The policy", p. 172. See also Sir William Armstrong, "Address of the President", *Proceedings of the Institution of Mechanical Engineers* (1861), 110–19, on p. 119.

[41] For a fuller discussion of this background, see Christine MacLeod, *Inventing the Industrial Revolution. The English Patent System, 1660–1800* (Cambridge: Cambridge University Press, 1988), chapter 11.

[42] William Derham, *Physico-theology: or, a Demonstration of the Being and Attributes of God from His Works of Creation* (London: W. Innys, 1713), pp. 306–9. For Derham, see W. Coleman, "Providence, capitalism and environmental degradation: English apologetics in an era of economic revolution", *Journal of the History of Ideas* (1976), **37**: 27–44, especially p. 32.

been part of God's preparations for the Reformation.[43] When Lord Camden, in 1774, railed against the notion of property in ideas, he still depicted successful inventors as the servants of Providence. "Those great men, those favoured mortals, those sublime spirits, who share that ray of divinity which we call genius, are entrusted by Providence with the delegated power of imparting to their fellow creatures that instruction which heaven meant for universal benefit".[44] Nineteenth-century commentators voiced a secularized Providence. In 1854 the words of an anonymous reviewer in *The Economist* suggest a conception which is more teleological than sociological: "Scientific discoveries and the arts built on them are neither made fortuitously nor by man's design; they are a regular and progressive development which no conduct of the human understanding — though care in this respect may make individuals good, knowing, and wise — could ever bring about".[45]

But the true inheritors of the determinist tradition were to be found among the political radicals, many of them sharing a Nonconformist religious background. Joseph Priestley, in his egalitarian providentialism, even had the temerity in 1767 to question Isaac Newton's claims to "genius", preferring a Baconian emphasis on the democratic pursuit of knowledge: talk of genius obscured the real nature of scientific discovery in which "patience and industry" were paramount and could be "equalled by many persons".[46] Robert Owen notoriously preached the influence of environment on character: "the character of man is formed *for* him and not *by* him".[47]

---

[43] [J. Asgill], *An Essay for the Press* (London: A. Baldwin, 1712), p. 4; [Matthew Tindal], *An Essay Concerning the Power of the Magistrate, and the Rights of Mankind, in Matters of Religion* (London: Andrew Bell, 1697), p. 184; [Matthew Tindal?], *A Letter to a Member of Parliament, shewing, that a Restraint Press is Inconsistent with the Protestant Religion, and Dangerous to the Liberties of the Nation* (London, 1698), pp. 11, 22. My thanks to Mark Goldie for these references.

[44] William Cobbett (ed.), *The Parliamentary History of England from the Earliest Period to the Year 1803*, vols. 13–36, (London: Longman & Co., 1812–20), vol. 17, col. 999. For Camden, see *Dictionary of National Biography*, "Charles Pratt, 1st Earl Camden (1714–1794)".

[45] *The Economist*, 16 September 1854, p. 1021.

[46] Simon Schaffer, "Priestley and the politics of spirit", in R.G.W. Anderson and C. Lawrence (eds.), *Science, Medicine and Dissent. Joseph Priestley (1733–1804)* (London: The Science Museum, 1987), p. 45; J.G. McEvoy, "Electricity, knowledge and the nature of progress in Priestley's thought", *The British Journal for the History of Science* (1979), **12**: 1–30, on p. 14; Yeo, "Genius, method, and morality", pp. 264–5.

[47] Quoted in Brinley Thomas, *The Industrial Revolution and the Atlantic Economy* (London: Routledge, 1993), p. 126. This was an axiom which Smiles explicitly repudiated, although he endorsed an anti-elitist notion of inventiveness. See Asa Briggs, *Victorian People* (Chicago: University of Chicago Press, 1970), p. 147; Yeo, "Genius, method, and morality", p. 267.

The most sustained account of invention which I have found in this period was by Thomas Hodgskin, the radical journalist who had a considerable influence on, among others, Karl Marx. Sharing with Priestley an egalitarian stance and a "belief in the existence of providential harmony in nature",[48] and with Owen the view that "every individual ... [is] fashioned by the time at which he lives, and by the society of which he is a member",[49] Hodgskin went on to elaborate a sophisticated rebuttal of inventive individualism and nascent heroism, in his *Popular Political Economy*, published in 1827. James Watt, argued Hodgskin, was simply born in the right time and place: his engine would have been "of no utility except in crowded countries, in which fuel is plentiful and manufactures established"; elsewhere "there would be no body to make or use it, no purpose to which it could be applied"; even in Britain a century earlier, both the incentive and the knowledge needed for its application would be lacking. Watt should be regarded, Hodgskin continued, simply "as one of those master-spirits who gather and concentrate within themselves some great but scattered truths, the consequences of numberless previous discoveries which, fortunately for them, are just dawning on society as they arrive at the age of reflection. They have the happy art to connect, by some little additional discovery of their own, the various truths lately brought into day".[50] The inventor was scarcely more than a puppet in the grand theatre of progress.

The extent to which the abolitionists campaigning in the third quarter of the nineteenth century drew on these earlier ideologies is unclear; I have found no direct references. However, deterministic explanations of invention, both secular and religious, were available to those, like Brunel, who for practical reasons wished to dispense with the patent system.

# THE HEROIC IDEOLOGY

For those who campaigned to retain the patent system, there were emerging alternative, individualistic explanations, as well, of course, as the individual attribution of inventions that the patent system itself had helped construct. By the late-eighteenth century, many of those who discussed invention were stressing the role of human creativity as an innate and individual attribute; the role played by God or Providence was increasingly limited to supplying the raw materials, including the laws of nature, from which the inventor would select in

---

[48] Elie Halevy, *Thomas Hodgskin*, trans A.J. Taylor (London: Ernest Benn, 1956), p. 36.
[49] Thomas Hodgskin, *Popular Political Economy. Four Lectures delivered at the London Mechanics' Institution* (London: printed for Charles Tait and William Tait, 1827), p. 87.
[50] Ibid., pp. 87–8.

order to construct an infinity of inventions.[51] This was, for example, the picture drawn by the engineer, Joseph Bramah, when helping to contest Watt's patent in 1797: God "stocked the universal storehouse ... out of which the same creating will directs every man to go and take materials, fit in kind and quality, for the execution of his design". The overall design might still be God's, but invention was overwhelmingly the product of "efforts of the mind and understanding" to synthesize "new effects from the varied applications of the same cause".[52] This inventor was no puppet.

From such "efforts of the mind" might spring "a thumping child of my brain", as James Nasmyth dubbed his steam hammer.[53] Such "products of the intellect" and "children of the brain" were deemed by many after 1850 to be peculiarly the property of their mental parents. "A man has as perfect a right to ownership in the production of his own intellect as to any other property which he might be possessed of", asserted Mr Macgregor, M.P., in 1851.[54] Eleven years later another M.P., Vincent Scully, leapt to the system's defence with the opinion that "there was nothing so peculiarly the property of a man as the labour of his brains".[55] This claim to a natural right in intellectual property had been little advocated in Britain, unlike in France.[56] It owed its new popularity at mid century to the patent controversy. It was complementary to the view that great inventions were the creation of genius, of singularly able and *irreplaceable* intellects. But invention could not be *too* easy — even for great minds. That would concede a major point to those abolitionists who argued that patents were not needed as an incentive since most inventors could not help but invent. The

---

[51] For similar views in the seventeenth and early eighteenth centuries, see MacLeod, *Inventing the Industrial Revolution*, pp. 203–4.

[52] Joseph Bramah, *A Letter to the Rt Hon Sir James Eyre, Lord Chief Justice of the Common Pleas; on the Subject of the Cause, Boulton & Watt v. Hornblower & Maberly: for Infringement on Mr Watt's Patent for an Improvement in the Steam engine* (London: John Stockdale, 1797), p. 77. See also Adam Smith, *Lectures on Jurisprudence*, ed. R.L. Meek, D.D. Raphael, and P.G. Stein (Oxford: Oxford University Press, 1978), p. 347; Dugald Stewart, *Lectures on Political Economy*, vol. 1, in Sir William Hamilton (ed.), *The Collected Works of Dugald Stewart* (11 vols., Edinburgh: Thomas Constable & Co.; London: Hamilton, Adams & Co., 1854–60), p. 196; *Observations on the Utility of Patents, and on the Sentiments of Lord Kenyon Respecting that Subject* (4th edn., London, 1791), p. 42.

[53] Samuel Smiles (ed.), *James Nasmyth, Engineer. An Autobiography* (London: John Murray, 1883), p. 247.

[54] *Hansard*, 3rd ser. CXVIII (1851), col. 1541.

[55] *Hansard*, 3rd ser. CLXVII (1862), col. 51. See also, A. Percy Sinnett, *Patent Rights. An Inquiry into their Nature* (London: James Ridgway, 1862), pp. 10–13.

[56] Dutton, *The Patent System*, pp. 17–18.

heroism of invention lay in the perseverance, the years of unremitting labour: this *combined* with a natural faculty for invention produced the great inventor.[57] It was such determination to succeed against the odds that explicitly marked out Smiles's heroes.[58]

It goes without saying that proponents of the heroic school denied demand-led or incremental explanations of invention. Henry Dircks, an engineer and author of the period's most sustained exposition of heroic views, *Inventors and Inventions* (1867), hopefully proclaimed: "That 'necessity is the mother of invention', is an aphorism that has long been exploded".[59] For Dircks, more elitist than Smiles, invention was totally unpredictable and often ahead of its time; only minor inventions or improvements occurred simultaneously. Inventiveness was a rare, personal quality: "Invention is not an art; it is a faculty of the mind more strongly developed in some men than in others, and is possessed by comparatively few in any great degree. No amount of acquired knowledge, no variety of scientific information, and no concentration of the mind on single subjects, would of themselves have any tendency to induce invention, in the absence of a natural ability that way".[60]

Possibly most difficult to refute was the disparaging charge made by Coryton, among others, that inventors merely applied the findings of scientific research. Responding directly to Coryton's paper in 1874, Thomas Webster, the patent agent, criticized him for "confusing discovery with invention.... The inventor takes the discovery and puts it to a practical use in the ordinary purposes of life. The labour of invention is immense".[61] Pushed by the abolitionists' arguments, their opponents followed through their own logic to produce more sophisticated accounts of invention. Some began to contend that there was more than one solution to every technical challenge and that good design had a role to play. John Howard, M.P. and agricultural machinery maker, defended the analogy between patents and copyright in 1869: "no two men ever wrote the same book at the same time, ... but he had frequently found that two authors writing almost simultaneously conveyed precisely the same ideas, though not in the same language. Exactly so, if two men invented similar machines simultaneously, it was never found they carried out their ideas precisely in the same mechanical way.... The inventor made use of the laws of

---

[57] For example, Sinnett, *Patent Rights*, p. 13; *Hansard*, 3rd ser. CXVIII (1851), cols. 1534–5, 1544; Thackray, "Natural knowledge", p. 687.

[58] For example, Samuel Smiles, *Men of Invention and Industry* (London: John Murray, 1884), pp. 57, 60, 72, 77. See also, Cockshut, *Truth to Life*, pp. 111–13.

[59] Henry Dircks, *Inventors and Inventions* (London: E. and F.N. Spon, 1867), p. 11. See also Woodcroft, *Brief Biographies*, p. 3.

[60] Dircks, *Inventors and Inventions*, pp. 7, 39–40.

[61] Coryton, "The policy", pp. 188–9.

nature just as the author of a book used the common language of mankind".[62]
Here was the individualism of creativity being given a new twist.

## THE LEGACY

One may speculate anachronistically that if Thomas Hodgskin had still been alive, he might have retorted that the differences were not representations of individuality but, rather, were shaped by the varying social environments and interests of their inventors. But, of course, he was not and he could not. The patent system survived; an heroic ideology of invention prevailed; and the debate fizzled out. As the patent system itself had helped construct an individualistic understanding of invention, so the patent controversy, in overwhelming the determinist explanation, affirmed and buttressed that construction.

Academic interest in the subject began to revive in the inter-war period, primarily among American sociologists and psychologists who emphasized "the status of culture", the evolutionary nature of technological change, and the frequent simultaneity of inventions, discounting the role of individual inventors. "Contrary to popular impression", wrote Ogburn and Thomas in 1922, "Watt, great man though he was, does not seem to have been indispensable to the perfection of the steam engine".[63] To little avail: an heroic ideology continued — and continues — to inform the popular conception of invention and technological change; when the word "inventor" is uttered, the image of Watt and his kettle probably springs to most British minds (and Edison, to most American). This we largely owe to the celebration of inventors stimulated by the patent controversy.

---

[62] *Hansard*, 3rd ser. CXCVI (1869), col. 912. See also ibid., cols. 920–1. For the current renewed emphasis on design, as the major feature distinguishing technology from science, see Eugene S. Ferguson, *Engineering and the Mind's Eye* (Cambridge, Mass.: M.I.T. Press, 1992); Walter G. Vincenti, *What Engineers Know and How They Know It. Analytical Studies from Aeronautical History* (Baltimore, Md.: Johns Hopkins University Press, 1990).

[63] Ogborn and Thomas, "Are inventions inevitable?", p. 91. For similar attacks on "the traditional point of view", see W.F. Ogborn, "The great man versus social forces", *Social Forces* (1926), **5**: 225–31 (225, 227); Gilfillan, *Inventing the Ship*, pp. ix, 3, 77–8.

# Technological Change During the First Industrial Revolution: The Paradigm Case of Textiles, 1688–1851

*Patrick O'Brien, Trevor Griffiths, and Philip Hunt*

## MODERN INTERPRETATIONS OF THE INDUSTRIAL REVOLUTION

In several ways, current reinterpretations of the first industrial revolution necessitate a reconsideration of the sources of technological progress in British textile production, particularly those "prototype" or "macro inventions" from John Kay's flying shuttle in 1733 to Edmund Cartwright's power loom in 1785. In the fullness of time, the classic inventions of this period led to the mechanization and relocation of production, accompanied by an extraordinary acceleration in the growth of output and a pronounced decline in the costs of producing cotton, woollen, linen, silk, and many mixed varieties of textiles. This "story", endlessly narrated, has never been explained by historians, who lack a general theory able to account for the major breakthroughs in technology that occurred in textiles over the eighteenth century.[1] Yet, paradoxically, the now hegemonic conception of the industrial revolution which plays down the role of new

---

[1] D.S. Landes, *The Unbound Prometheus. Technological Change and Industrial Development in Western Europe from 1750 to the Present* (Cambridge: Cambridge University Press, 1969), pp. 80–88; T.S. Ashton, *The Industrial Revolution, 1760–1830* (London: Oxford University Press, 1948), pp. 58–93, 216; N. Rosenberg, *Inside the Black Box. Technology and Economics* (Cambridge: Cambridge University Press, 1982), pp. 4–27.

technology and which has encouraged polemical attempts to expunge the whole notion of an industrial revolution from "sensible historical discourse" has made the task of formulating a heuristic theory more, rather than less, urgent.[2]

Revived interest in Britain's famous transition to industrial society has taken the form of responses and rejoinders to the research of quantifiers working within orthodox frameworks of macro-economic analysis. This cliometric approach, based upon imperfect data, moulded and calibrated into index numbers, social accounts, partial productivity indicators, and above all into a growth accounting framework, in order to relate increases in national output to long-term changes in the inputs of land, labour, and capital formation, need not be unpacked here. But its hypotheses are clear enough.

Firstly, on all the indicators that have been constructed to measure the pace and pattern of British economic growth between, say, 1688 and 1851, the first industrial revolution turns out to have been a so much slower and less dramatic discontinuity that several historians have dismissed it as a myth or a misnamed episode in European economic and technological history.[3] When we consider the reconstituted data available to measure the pace of economic change for Britain as a whole (i.e. growth rates in real per capita income, industrial output per head, and in the productivity of labour employed in manufacturing and agriculture) then the industrial revolution as a widely diffused national event does not come on stream until well into the nineteenth century, which is several decades later than Ashton's famous narrative and Deane and Cole's initial and influential attempts at quantification suggested.[4] Nevertheless, if we compare estimated rates of change for the first half of the eighteenth century with those for the second quarter of the nineteenth century, marked discontinuities, especially in industrial output, remain inescapable. Furthermore, contemporaries knew that Britain's economy and society had undergone profound changes which appeared set to continue at an accelerated rate. Foreigners who visited the Hanoverian kingdom after Waterloo concurred. They recognized a successful industrial economy when they saw one and knew their own countries would have to "catch up".[5]

---

[2] P.K. O'Brien, "Introduction: modern conceptions of the industrial revolution", in P.K. O'Brien and R. Quinault (eds.), *The Industrial Revolution and British Society* (Cambridge: Cambridge University Press, 1993), pp. 1–30.

[3] M. Berg and P. Hudson, "Rehabilitating the industrial revolution", *Economic History Review* (1992), 2nd ser. **45**: 24–50; N.F.R. Crafts and C.K. Harley, "Output growth and the British industrial revolution: a restatement of the Crafts–Harley view", *Economic History Review* (1992), 2nd ser. **45**: 703–30.

[4] Ashton, *Industrial Revolution*; P.M. Deane and W.A. Cole, *British Economic Growth, 1688–1959. Trends and Structure* (Cambridge: Cambridge University Press, 1962).

[5] P.K. O'Brien and C. Keyder, *Economic Growth in Britain and France 1780–1914. Two Paths to the Twentieth Century* (London: Allen and Unwin, 1978), pp. 186–8.

Secondly, the initial phases of the industrial revolution are now depicted by cliometricians as "extensive" rather than "intensive": meaning that very high proportions of the increments to national output and to industrial production before 1825, can be attributed to the employment of more labour and additional inputs of capital rather than to increases in factor productivity, i.e. higher outputs per man hour or per unit of capital employed.

Thirdly, the new cliometric interpretation observes that where and when productivity improvements occurred, they tended to be located in just a few sectors of the economy. Within industry, the influence of breakthroughs and improvements in technology before the second quarter of the nineteenth century were confined to basic metallurgy and to textiles, above all to cotton fabrics. As late as the 1830s, the mechanization of cotton production and the concentration of all the processes involved in the preparation, spinning, weaving, and finishing of cotton cloth into steam-powered factories located in towns, represented a paradigm for other industries to emulate.[6] In this sense, the industrial revolution is currently presented as an example of "unbalanced growth".[7]

In many ways, this interpretation of the first industrial revolution reads like a very old story in which textiles in general, and cotton in particular, are taken as exemplary cases of early mechanization.[8] Of course, the story is highly contested, especially by historians whose research on regions, proto-industrialization, and transformations in the organization of traditional work processes lead them to construct narratives illustrating a more broadly based sequence of technical change.[9] Debate continues, but for the purposes of this essay the cliometric painting of the industrial revolution provides historiographical justification for focussing again on textiles and upon the sequence of famous machines, processes, and improvements that emerged between the times of Kay and Cartwright. Since so much of the early action took place within this industry and was concentrated within a period of six or seven decades in the eighteenth century, subsequent, more rapid, and pervasive industrial growth in the Regency and Victorian periods can be depicted as an elaboration of technological knowledge brought to maturity much earlier and within the confines of a single industry, albeit one that accounted for a large share of manufacturing activity in Britain. Thus if historians could offer some kind of general explanation for the technological innovations and improvements that transformed the

---

[6] R. Samuel, "Workshop of the world: steam power and hand technology in mid-Victorian Britain", *History Workshop* (1977), **3**: 6–72;   P. Hudson, *The Industrial Revolution* (London: Edward Arnold, 1992), pp. 21–4.

[7] N.F.R. Crafts, *British Economic Growth During the Industrial Revolution* (Oxford: Clarendon Press, 1985).

[8] P. Mantoux, *The Industrial Revolution in the Eighteenth Century. An Outline of the Beginnings of the Factory System in England* (1928; London: Methuen, 1964).

[9] Berg and Hudson, "Rehabilitating the industrial revolution", pp. 24–50.

making of cloth over the century following the introduction of Kay's flying shuttle in 1733, they might be on the way to communicating an understanding of the prime mover behind British and European industrialization before the introduction of a new set of industries and technologies in the late nineteenth and early twentieth centuries.[10]

# THEORIES OF TECHNICAL CHANGE

The proper historical context for the study of technological discoveries is local and specific because innovations consist of: (a) new products or variations on old products sold to consumers, or (b) artefacts or processes designed to raise the quality of commodities produced, while holding the overall cost of the inputs used in their production constant, or, most commonly of all, (c) the techniques that lowered costs of production by reducing the quantities of capital, labour, time, raw materials, energy, etc. per unit of output. Nevertheless, most historians treat innovations as particular products, artefacts, or processes, and they study them and aspire to communicate their findings at readable and memorable levels of generalization. Our research into the origins of the myriad of techniques (inventions, improvements, novel products) that carried textile production in Britain forward from 1688 to the scale and level of international competitive efficiency that it had so clearly reached by 1851, is predicated on the assumption that we shall offer some interesting and new generalizations about the forces behind the rise in productivity for a major sector of national industrial output.

Textile innovations comprehended the new techniques and processes that were introduced to improve the transformation of four agricultural raw materials (wool, flax, silk, cotton, and mixes of those natural fibres) into finished (i.e. bleached, dyed, and printed) cloth. Between 1688 and 1851, the list of innovations would have been extremely long, and only a fraction of the entire flow of improved technology introduced over time is now recoverable from the sources, which include: published accounts of the famous machines; patent specifications; industrial and business histories; and other contemporary sources. Nevertheless, countless improvements that lowered costs of production and differentiated or raised the quality of cloth also occurred, but these are now, alas, lost to historians seeking to reconstruct the sequence of technological change evolving through long periods. Working upon the surviving data, available for patented and non-patented textile innovations alike, familiar taxonomies have been imposed in order to divide "product" from "process" innovations and to differentiate process innovations into four categories: preparation, spinning, weaving, and finishing. More problematically, macro inventions have been

---

[10] M. Dintenfass, *The Decline of Industrial Britain 1870–1980* (London: Routledge, 1992).

distinguished from improve-ments concerned to adapt a machine or process or product, in order to bring it into efficient day-to-day use and production for the market. As we study the series of major inventions and improvements, which ultimately revolutionized all four processes involved in the cloth manufacture, it is clear that they appeared discontinuously, even haphazardly, over a truncated period. At the end of this sequence (let us date it to about the time of the Great Exhibition of 1851) cloth production had been transformed from a handicraft proto-manufacture, using some machinery and water power, into a mechanized, steam-powered, factory-based urban industry.[11] In the context of millennia since craftsmen and women had been engaged in making cloth, this period of radical change is so short, the rate of transformation within and across all four stages of production so rapid, and the original locus so geographically concentrated in one kingdom, that the "British" revolution in textiles has been deservedly recognized as a seminal episode in the history of technology.

But can historians explain that famous episode in a satisfying, general, and communicable way? Not, we suggest, by dealing in narrative form with one innovation after another, although that kind of tightly focussed research is a precondition for the interplay of fact and concept required to theorize about technological change. Nevertheless, historical narratives based upon a rounded comprehension of the context within which particular artefacts, processes, and products first emerged certainly offer more persuasive explanations for tech-nological progress during the industrial revolution than does recourse to those overarching theories drawn from economics or sociology which attempt, unsuc-cessfully, to account for the pace and pattern of technological discovery.[12]

For example, objections may be raised against both demand and supply-side theories of invention where the inducements to invent flow from consumer needs and expenditures or from shortages or bottlenecks experienced by pro-ducers. Once the cloth industries of Western Europe became engaged in competitive production for home and foreign markets, potentially profitable inventions were always in demand. Furthermore, growing shortages of labour, raw materials, and other inputs were neither sufficient nor necessary to prompt merchants and businessmen actively to seek or tacitly to encourage the search

---

[11] I. Inkster, *Science and Technology in History. An Approach to Industrial Development* (Basingstoke: Macmillan, 1991), pp. 1–13, 32–88; C. Singer, E.J. Holmyard, A.R. Hall, and T.I. Williams (eds.), *A History of Technology. Vol. III. From the Renaissance to the Industrial Revolution c.1500–c.1750* (Oxford: Clarendon Press, 1957), pp. 151–205; C. Singer, E.J. Holmyard, A.R. Hall, and T.I. Williams (eds.), *A History of Technology. Vol. IV. The Industrial Revolution c.1750–c.1880* (Oxford: Clarendon Press, 1988), pp. 230–57, 277–327.

[12] Rosenberg, *Inside the Black Box*, pp. 4–27; W.E. Bijker, T.P. Hughes, and T.J. Pinch (eds.), *The Social Construction of Technological Systems. New Directions in the Sociology and History of Technology* (Cambridge, Mass.: MIT Press, 1987).

for innovation that might differentiate their products or lower their costs of production.[13]

In any case, as specified, neither demand nor supply theories can possibly be tested for the relevant decades that preceded and witnessed the rapid transformation of the British textile industry. For example, were demand "pressures" emanating from consumers for cheaper and/or novel forms of cloth greater on British textile producers than on their Dutch and French rivals? Did such demands intensify prior to the acceleration in the pace of technological innovation that is suggested, for example, by the upswing in numbers of patents taken out in London in the 1760s? Why did the spinning jenny and the water frame appear in that decade and not before, when presumably the level of demand for cheaper yarn was equally buoyant? If demand-pull were a dominating influence on the pace and timing of technological progress, why did it take so long to perfect and diffuse most of the famous mechanical breakthroughs in weaving by Kay, Cartwright, and others?

Latterly, writings on the rise of material culture or the consumer revolution in the seventeenth and eighteenth centuries have led to a revival of demand-led theories of industrial growth and innovation. Histories of consumer behaviour have been concerned to argue the hypothesis that from the Restoration onwards, social, cultural, and political changes combined to alter propensities to consume in England, so that the home market became altogether more hospitable to merchants and industrialists trying to persuade households to buy a greater variety and volume of textiles and other wares.[14] The new cultural history of European materialism is properly concerned to emphasize that there is a demand side to economic (and *ipso facto* to technological) progress and that there is more to demand than falling prices and rising personal incomes. Economic growth requires populations not only able but also willing, and preferably eager,

---

[13] P.K. O'Brien, "The mainsprings of technological progress in western Europe 1750–1850", in P. Mathias and J.A. Davis (eds.), *Innovation and Technology in Europe from the Eighteenth Century to the Present Day* (Oxford: Blackwell, 1991), pp. 6–17; J. Mokyr, *The Lever of Riches. Technological Creativity and Economic Progress* (New York and Oxford: Oxford University Press, 1990), pp. 57–112, 151–92, 239–300.

[14] C. Mukerji, *From Graven Images. Patterns of Modern Materialism* (New York: Columbia University Press, 1983), pp. 166–261; C. Campbell, *The Romantic Ethic and the Spirit of Modern Consumerism* (Oxford: Blackwell, 1987); N. McKendrick, J. Brewer, and J.H. Plumb, *The Birth of a Consumer Society. The Commercialization of Eighteenth Century England* (London: Europa, 1982); J. Mokyr, "Demand vs. supply in the industrial revolution", *Journal of Economic History* (1977), **37**: 981–1008; A.Y.B. Schachar, "Demand vs. supply in the industrial revolution: a comment", *Journal of Economic History* (1984), **44**: 801–5; E. Gilboy, "Demand as a factor in the industrial revolution", in R.M. Hartwell (ed.), *The Causes of the Industrial Revolution in England* (London: Methuen, 1967), pp. 121–38.

to consume the products of capitalist industry. In turn, that willingness depends upon the proclivities of households to: admit novel material goods into private domains; convert leisure into work in order to spend on the "superfluities" of the day; sustain levels of consumption in the face of adverse movements in their real incomes; emulate the consumption patterns of neighbours and betters; and display and fashion their identities through conspicuous consumption. Such propensities are culturally ordered and change more or less slowly through time.[15]

The difficulty with the "rise of material culture" thesis is that it seems almost impossible to locate changes that can be isolated as peculiarly British or to date discontinuities in consumer behaviour that were powerful enough to sustain the pressure of demand at the levels required to promote a continuous flow of innovation. The thesis is not without a place in the story of the industrial revolution, but compared with several *supply-side* forces operating to widen domestic, imperial, and foreign markets for textiles, it is difficult to accord changes in culture, themselves in part economically conditioned, much "autonomous" weight. In point of time, the posited rise of material culture was probably not bounded by national frontiers, and it certainly coincided with: the integration of the market through investment in transportation and improved networks for the distribution of textiles of all kinds; changes in imperial and foreign policy which secured markets overseas; the growth in agricultural productivity which, in the face of an upswing in population, restrained the proportion of income that households devoted to expenditure upon basic foodstuffs; rising rates of investment in urban construction and industry; and, above all, the highly conscious attempts by producers, merchants, and finishers of cloth actively to cajole consumers into buying more through an entirely traditional process of product differentiation. Cultural changes might well be important and even autonomous in their impact on demand, but they need to be specified, dated, and related more clearly to measured changes in textile production.[16]

Unfortunately supply side or inducement models are no less difficult to link in any convincing way to the pace and pattern of technological discovery. Although the mercantilist literature of the period is replete with observations about the high level of English, compared with French and Irish, wages and with complaints about the idleness, insubordination, and irrational preference for leisure displayed by English workmen, the wage data required to test the

---

[15] C. Shammas, *The Pre-industrial Consumer in England and America* (Oxford: Clarendon Press, 1990); B. Lemire, *Fashion's Favourite. The Cotton Trade and The Consumer in Britain, 1660–1800* (Oxford: Oxford University Press, 1991). See also J. Brewer and R.S. Porter (eds.), *Consumption and the World of Goods* (London: Routledge, 1993).

[16] T. Griffiths, P.A. Hunt, and P.K. O'Brien, "Inventive activity in the British textile industry, 1700–1800", *Journal of Economic History* (1992), **52**: 881–906.

suggestion that the incentives for British textile producers to promote a search for labour-saving machinery became more powerful over the eighteenth century, are simply not there.[17] At the same time as commentators and politicians called for policies to put pauper, child, and female labour to work at spinning wheels, worries were being expressed in Anglican and mercantilist literature about under-employment and unemployment among adult males, particularly after population growth accelerated in the second half of the eighteenth century.[18]

Firstly and at national level, the case for increasingly inelastic supplies of labour available for work in industry is not compelling. English population growth accelerated earlier and more rapidly than other European populations. Secondly, from mid-century onwards, food prices began their upward climb and presumably compelled the previously idle to work harder in order to maintain consumption levels for themselves and their families.[19] Thirdly, labour market demarcations based upon gender and traditional skills broke down, enabling "reserve armies" of women, children, and Celts to move into industrial employment. Finally, textile innovations seem to have diffused more readily in relatively low wage regions (e.g. in Lancashire before Wiltshire).[20]

Perhaps the only way around the present impasse faced by economic historians likely to remain ignorant about eighteenth-century labour markets in general and movements in real product wages in particular, is more research which attempts to reconstruct local matrices of demand and supply for the rather specialized labour required to produce the astonishing multiplicity of fabrics that made up the nation's aggregate textile output. Neither the final output, cloth, nor the labour and skills required to treat, spin, and weave natural fibres were homogeneous. At present, the tiny selection of case studies in print suggests that

---

[17] After an extensive search of primary and secondary sources on wages in the eighteenth century, we have reluctantly concluded that nothing definite can be said about eighteenth-century textile wages rates.

[18] For worries about under-employment, see T.W. Hutchison, *Before Adam Smith. The Emergence of Political Economy 1662–1776* (Oxford: Blackwell, 1988); T.E. Gregory, "The economics of employment in England, 1660–1713", *Economica* (1921), **1**: 40–4; A.W. Coats, "Changing attitudes to labour in the mid-eighteenth century", *Economic History Review* (1958–9), 2nd ser. **11**: 35–51; A.W. Coats, "The relief of poverty: attitudes to labour and economic change in England 1660–1782", *International Review of Social History* (1976), **21**: 108–9.

[19] P.K. O'Brien, "Agriculture and the home market for English industry, 1660–1820", *English Historical Review* (1985), **100**: 773–99.

[20] E.H. Hunt and F.W. Botham, "Wages in Britain during the industrial revolution", *Economic History Review* (1987), 2nd ser. **40**: 380–99; E.H. Hunt, "Industrialization and regional inequality: wages in Britain, 1760–1914", *Journal of Economic History* (1986), **46**: 935–66, especially p. 952.

the impetus to innovation, even to process innovation, was as often influenced by entrepreneurial concerns to upgrade quality and develop contingent skills among local workforces as by an urgent and peculiarly British need to circumvent any rise in real product wages.[21]

Other familiar inducement mechanisms are equally difficult to validate, especially the ever-popular challenge and response model. Textbook accounts which repeat that model explain the sequence and timing of innovations across the four major stages of cloth production in terms of imbalances, whereby the diffusion of a new productive technique in one stage sets up pressures for a response both up but particularly downstream, to deal with intensified demands for inputs or more elastic and cheaper supplies of outputs. The most frequently cited example is Kay's shuttle, which improved the productivity of weavers whose enhanced demand for yarn was only satisfied, in the fullness of time, by major spinning inventions from Hargreaves, Arkwright, and Crompton. In turn, the new spinning machines produced a surfeit of yarn which prompted a search for power looms, eventually resolved after an unexplained "lag" of some five decades by Roberts in 1822.[22] Apart from the often protracted lapses of time between challenge and response, which require explanation, there is no evidence in the statistics that we have gathered, related to patented and non-patented inventions alike, that innovation clustered around particular stages of production over well demarcated periods of time. Secondly, Kay's shuttle, which plays such a pivotal role in the story, cannot be explained either in terms of a downstream response to prior inventions in the finishing of cloth or as induced by the greater accessibility of cheaper supplies of yarn. Furthermore, Kay's invention was simply too circumscribed in its effects to be linked convincingly to the "wave" of significant innovations and improvements in spinning, associated with the

---

[21] E. Richards, "Women in the British economy since about 1700: an interpretation", *History* (1974), **59**: 337–57; D. Bythell, *The Handloom Weavers. A Study in the English Cotton Industry During the Industrial Revolution* (Cambridge: Cambridge Univeristy Press, 1969), pp. 42–65; E. Kerridge, *Textile Manufactures in Early Modern England* (Manchester: Manchester University Press, 1985), pp. 235–8; E. Baines, *Observations on Woollen Machinery* (Leeds, 1803); J. Anstie, *Observations on the Importance and Necessity of Introducing Improved Machinery into the Woollen Manufactory* (n.p., 1803).

[22] Landes, *Unbound Prometheus*, pp. 41–88. For "challenge and response", see P. Mantoux, *The Industrial Revolution in the Eighteenth Century* (1964 edn.), pp. 208–9, 239; T. Sutcliffe, *An Exposition of Facts Relating to the Rise and Progress of the Woollen, Linen and Cotton Manufactures of Great Britain* (Manchester, 1843); and M.M. Edwards, *The Growth of the British Cotton Trade 1780–1815* (Manchester: Manchester University Press, 1967), p. 3.

names of that celebrated Lancashire trio Hargreaves, Arkwright, and Crompton.[23]

In our view, the proper historical context for focussed exploration into the "origins" of the jenny, water-frame, and mule resides in political economy: firstly, in the substitution of domestically produced cotton for Indian calicoes on the home market, and, secondly, in a politically sponsored development of linen cloth manufactured in Ireland, which created instabilities and shortages in the supply of Irish linen yarn exports to Lancashire for the production of fustians.[24] We would argue that historians of the justly acclaimed cluster of breakthroughs in spinning in Lancashire should bring into their stories the Irish, and African, connexions with the fustian and proto-cotton industry that had developed in that county long before the Seven Years War. Bourdieu's "habitus", as a product of history producing individual and collective practices that are recognizable without being strictly predictable, seems to "fit" the situation in Lancashire in the eighteenth century and to account for the spinning machines of its famous inventors rather well.[25] At least, Bourdieu's modest concept seems to have more to offer historians who are concerned about the interplay of fact, concept, and vocabulary, and who wish to generalize about technological discovery, than does recourse to demand-led, supply-induced, and socially constructed theories of invention.

It is unfortunate that while historians are rather effective at exposing fault lines in the imported theoretical assumptions behind text-book accounts of the

---

[23] For the limited effects of Kay's shuttle on narrow goods, described in Kay's own submission to the Society for the Encouragement of Arts (later the Royal Society of Arts), see A. Paulinyi, "John Kay's flying shuttle: some consideration on its technical capacity and economic impact", *Textile History* (1986), **17**: 149–66, on p. 154; and Sir H.T. Wood, "The inventions of John Kay, 1704–70", *Journal of the Royal Society of Arts* (1911–12), **60**: 73–86, on pp. 83–4.

[24] Fustians were cloths made of linen warp and cotton weft. P.K. O'Brien, T. Griffiths, and P. Hunt, "Political components of the Industrial Revolution: Parliament and the English cotton textile industry, 1660–1774", *Economic History Review* (1991), 2nd ser. **44**: 394–423. We are exploring the links between imported Irish linen yarn and innovations in spinning machines in Lancashire. For the Irish linen manufacture, see C. Gill, *The Rise of the Irish Linen Industry* (Oxford: Oxford University Press, 1925), pp. 61–81; L.M. Cullen, *Anglo-Irish Trade, 1660–1800* (Manchester: Manchester University Press, 1968), pp. 58–61. For the effects on English industry, the connexion with Ireland is made explicit in petitions to Parliament, e.g. *Journals of the House of Commons* XXV (London, 1811), p. 870; and in evidence to Commons Committees. See especially *Reports from the Committees of the House of Commons* ii, Miscellaneous subjects, 1738–65; *Report for the Committee appointed to examine and state to the House, the matter of fact in the several petitions of the manufacturers of, and traders and dealers in, the linen manufactory, reported by Lord Strange on 26 April 1751.*

[25] P. Bourdieu, *The Logic of Practice*, trans. Richard Nice (Cambridge: Polity, 1990).

pace and pattern of technological change during the industrial revolution, they remain less adept at synthesizing the considerable amount of printed case studies relevant to the subject of this paper. Overwhelmed with densely footnoted information for one innovation after another, the temptation to reduce the complex evolution of British textile production to a chronological narrative remains strong.

## THE EVIDENCE OF PATENT SPECIFICATIONS

Nevertheless, two related sources — patent specifications and the biographical details of the community of men (and a few women) who claimed to be, and in many cases were recognized as, successful innovators — might help us to construct some acceptable middle-range generalizations about technological change in textiles as a whole. To this end, we have established two data-bases: one designed to provide some sort of empirical foundation for an analysis of the pace and pattern of technological change, the other to explore the social, religious, educational, and cultural milieux of nearly 2500 inventors. Our programme of empirical research is still incomplete, and the inferences and suggestions in this paper will be presented more fully in a forthcoming monograph. Nevertheless, a preliminary summary from the findings seems possible.

First, we deal with evidence culled from patent statistics and specifications and other sources. In their quantitative work, cliometricians and economists are prone to aggregate recorded innovations into an index, purporting to represent annual and cyclical variations in the volume of technological change within particular industries or across national economies as a whole. Such an index would be extremely useful to historians, but, except for entirely limited purposes, no such indicator can be constructed, since innovations recorded in patents and other documents are an unknown and potentially variable proportion of changes in the total flow of innovation.[26] Even recorded innovations cannot be aggregated without some system of weighting to account for variations in their economic and technological significance. If this were not done, Hargreaves's spinning jenny would be accorded the same weight in the aggregate as Peter Vallotton's patent for manufacturing hosiery pieces adapted for the wear of persons afflicted with rheumatism and gout, which was taken out in the same year. Finally, changes in the propensity to patent and in the

---

[26] Griffiths, Hunt, and O'Brien, "Inventive activity", pp. 881–906.

commercial viability of patented inventions could seriously compromise the comparability of recorded totals in different periods.[27]

So no useful index of technological change can be constructed from a simple aggregation of patents, but within patent and other specifications of inventions, there is information that can be tabulated to expose changes in "patterns" of inventive activity as they evolved through time. For example, the urban and regional concentration of such activity is traceable in a rough and ready way through the recorded addresses of patentees. Patent specifications make reference to particular industries, to final products, and to specific machines, devices, and processes, and they include claims for the economic and social benefits which the patentees expected to accrue from their inventions.

For reasons adumbrated above, patents cannot be aggregated into an index that allows changes in the volume of innovation to be regressed upon other economic indicators in order to elucidate the correlated factors influencing innovation within a macro-economic framework.[28] Nevertheless, changes in the overall *scale* of innovation might be captured over long periods of time. The total numbers of patents registered from 1675 to 1850 changed as indicated in the Table.

As Ashton suggested a long time ago, these data can be reasonably interpreted as evidence that the volume of innovation increased discernibly over the long run and became a pervasive component behind British economic growth.[29] Other kinds of more precise correlations and inferences based upon finely tuned dis-continuities or upswings that appear on graphs (which plot numbers of patents recorded each year) are, to repeat, impossible to substantiate.[30]

---

[27] For a recent discussion, see Z. Griliches, "Patent statistics as economic indicators: a survey", *Journal of Economic Literature* (1990), **28**: 1661–1707; J. Schmookler, *Invention and Economic Growth* (Cambridge, Mass.: Harvard University Press, 1966); J. Schmookler, "Comment on Barkev S. Sanders, 'Some difficulties in measuring inventive activity'", in *National Bureau of Economic Research. The Rate and Direction of Inventive Activity: Economic and Social Factors* (Princeton, NJ: Princeton University Press, 1962), pp. 78–83; K.L. Sokoloff, "Inventive activity in early industrial America: evidence from patent records, 1790–1846", *Journal of Economic History* (1988), **48**: 813–50; R.J. Sullivan, "England's 'Age of Invention': the acceleration of patents and patentable invention during the Industrial Revolution", *Explorations in Economic History* (1989), **26**: 424–52; R.J. Sullivan, "The revolution of ideas: widespread patenting and invention during the English Industrial Revolution", *Journal of Economic History* (1990), **50**: 349–62; C. Macleod, *Inventing the Industrial Revolution. The English Patent System, 1660–1800* (Cambridge: Cambridge University Press, 1988) pp. 2–7, 115, 144–57.

[28] Griffiths, Hunt, and O'Brien, "Inventive activity", pp. 881–906.

[29] T.S. Ashton, "Some Statistics of the Industrial Revolution", *The Manchester School of Economic and Social Studies* (May 1948), **16**: 214–34.

[30] But for a contrary view, see Sullivan, "The revolution of ideas", pp. 349–62; and Sullivan, "England's 'Age of Invention'", pp. 424–52.

**Table:**   Numbers of Inventions and Textile Innovations Patented, 1675–1849:

|         | Total Patents | Textile Patents |
|---------|---------------|-----------------|
| 1675–99 | 187           | 34              |
| 1700–24 | 109           | 18              |
| 1725–49 | 178           | 34              |
| 1750–74 | 442           | 75              |
| 1775–99 | 1,273         | 236             |
| 1800–24 | 2,697         | 458             |
| 1825–49 | 7,848         | 1,509           |

*Sources:* B. Woodcroft (ed.), *Titles of Patents of Invention, Chronologically Arranged from March 2, 1617 (14 James I) to October 1, 1852 (16 Victoriae)* (2 vols., London, 1854); A.A. Gomme, "Date Corrections to English Patents, 1617–1852", *Transactions of the Newcomen Society* (1932–3), **13**; D.R. Jamieson, "Introduction", in B. Woodcroft (ed.), *Alphabetical Index of Patentees of Inventions* (London, 1854; reprinted London, 1969).

Nevertheless, other plausible inferences have been drawn from patent statistics. For example, the proportion of "professional" patentees — defined by Dutton as men who registered, say, two or more innovations and whose claims to intellectual property rights spanned more than a single industry — increased from 28 per cent in the period 1751–60 to 50 per cent between 1841 and 1850.[31]

Within textiles, which accounts for far and away the largest proportion of patents registered for any single industry between 1675 and 1849, the same long-term trends towards the specialization and professionalization of inventive activity can be detected. For example, there was a predictable tendency for inventive activity to emanate from the new and rising regions of cloth production in the midlands and north of England. Although the addresses of agents and businessmen "representing" genuine inventors complicate the geographical picture, the relative decline of London, East Anglia, and the southern counties as centres of inventive activity becomes transparent, especially after the turn of the century. Competitive pressures forced most patentees to insist that their innovations were not fabric specific, but could apply equally well to the manufacture of woollen, silk, linen, or cotton cloth and to mixes of those four fibres. Nevertheless, the rise of specialist manufacturers and machine-makers ensured that the whole process of innovation became much more endogenous to particular sections of the textile industry producing differentiated varieties of cloth, located within rather well defined manufacturing regions.

After the last war with France ended in 1815, the technological transformation of textile production proceeded in altogether more explicable ways. From around that time, innovation became increasingly dominated by improvements to

---

[31] H.I. Dutton, *The Patent System and Inventive Activity During the Industrial Revolution, 1750–1852* (Manchester: Manchester University Press, 1984), p. 114.

machines, processes, and products that had emerged in prototype form decades before. Machines, finishing processes, and preparatory techniques diffused from fabric to fabric. The process became more obviously led by successful local examples, as home and foreign markets generated price signals which induced a general and sustained search for cost saving and product innovations.

## THE MACRO TEXTILE INVENTIONS OF THE EIGHTEENTH CENTURY

If, as we tentatively suggest, the technological history of textile production might be divided into an age of discovery, followed in the nineteenth century by several decades when the implications of prototype production techniques and modes of organization worked their way through the several processes involved in the manufacture of four major fibres from wool, flax, silk, and cotton into bleached, dyed, and printed cloth, then the most interesting and difficult problem to explain is why so many of the prototype innovations happened to be British in their origin and early development.

The answer to that large question should, however, be prefaced with the obvious point that national economies which include industries that have reached certain minimal scales of production and maturity are more likely to generate innovations than countries with infant industries. By the early eighteenth century, the British Isles enjoyed a long tradition of comparative advantage in the production of most varieties of woollen cloth. Silk weaving and finishing had been established in London, the midlands, and Kent. Coarse qualities of linen subsidized by government were manufactured in Ireland and Scotland, while Lancashire specialized in fustians. British textile production included all major varieties of cloth as well as mixes of fibres finished and sold in an astonishing range of varieties and qualities. The textile industry as a whole had achieved a level of production and technical sophistication from which it could emulate most varieties of foreign cloth and yarn, absorb new technology, and even generate a flow of indigenous innovations.[32]

For something like a century after the Restoration, the kind of "technical progress" that carried British industry forward to ever higher and more variegated levels of output consisted overwhelmingly of product innovations — new mixes of yarns woven into cloth that was bleached, dyed, and printed in novel ways appealing to consumers at home and abroad. That pattern of innovation seems entirely traditional; it represented both a response to and a force behind the widening of domestic and foreign markets. Smithian growth (as it is

---

[32] C.G.A. Clay, *Economic Expansion and Social Change. England, 1500–1700* (2 vols., Cambridge: Cambridge University Press, 1984); D.C. Coleman, *The Economy of England, 1450–1750* (London: Oxford University Press, 1977); Kerridge, *Textile Manufactures*; E. Baines, *History of the Cotton Manufacture* (London, 1835).

now called) led the development of textile production within the kingdom upward to a plateau from where indigenous technological discoveries became more and more likely.[33]

Fundamental breakthroughs (macro inventions) can be observed in the emergence of new spinning machines that appeared between 1764 and 1779, – power looms appearing between 1785 and 1822, and machinery for preparatory processes and a range of chemical techniques and mechanical devices for finishing cloth emerging between 1783 and 1800. With the exception of some new and important processes involved in the bleaching and dyeing of cloth, almost the entire spectrum of advanced technology which ultimately transformed the manufacture of textiles was initially developed by British inventors and improved by British artisans to the point where it could be exploited commercially by British businessmen. With hindsight, historians can now plausibly suggest that the British textile industry was able to absorb, sponsor, promote, and encourage a search for improved technology at any time during the eighteenth century. However, they remain at a loss to account for the astonishing range of prototype discoveries (power-driven machines, new techniques for preparation and finishing, and factory modes of organization, as well as the continued proliferation of differentiated varieties of cloth) that came on stream over such a truncated period of time in the long history of this global industry.

Our approach to this difficult but central problem in technological and economic history is first to reiterate the familiar observation that the critical early phases of process innovation in textiles were heavily concentrated around one fabric, namely cotton, and that technologies, techniques, and processes which developed in relation to that particular fibre were subsequently adapted and diffused across the stages and processes for the manufacture of woollen and linen cloth. Of course, this statement deliberately simplifies a complex process of interaction across the separate and, in some respects, technologically autonomous branches of textile production. Nevertheless, it serves as a heuristic generalization that carries the discussion forward to address questions of why cotton and why cotton in Britain?

The tensile properties of cotton fibres that made them peculiarly amenable to mechanized spinning into yarns capable of withstanding the pressures involved in weaving on power-driven looms have already been carefully analysed in the literature. Cotton cloth also proved to be more adaptable than fabrics woven from other fibres to the new finishing techniques for bleaching, printing and dyeing that appeared at the end of the eighteenth century.[34] Furthermore, the price of the basic raw material produced by slave labour in the Americas declined in relation to the costs of hemp, flax, wool, and silk produced by cheap but relatively "free" agricultural labour in Europe. Hence incentives to allocate resources and to search

---

[33] Griffiths, Hunt, and O'Brien, "Inventive activity", pp. 881–906.

[34] Baines, *History of the Cotton Manufacture*; A.P. Wadsworth and J. de L. Mann, *The Cotton Trade and Industrial Lancashire 1600–1780* (Manchester: Manchester University Press, 1931); Edwards, *Growth of the British Cotton Trade*.

for new varieties of cloth and improved technologies that used cotton fibres intensified over the eighteenth century as elastic and relatively low-cost supplies of raw cotton from the New World arrived in European ports.[35]

More for political than for economic reasons, British textile producers became, from an early stage, steadily more enthusiastic in their use of cotton fibres than their rivals on the continent. British colonies in the Caribbean and plantations on the American mainland grew cotton, but the major impetus behind the rapid development of a domestic market for cotton cloth came from trade with India. Between 1660 and 1700, rapidly increasing imports of Indian cotton cloth, especially in printed form, demonstrated the potential size of the British home market for the "new" calicoes, muslins and nanqueens. Under pressure from established textile manufacturing lobbies, Parliament (in contrast with governments on the continent) moved first to curtail and then to exclude Asian cottons from home markets. By the time such bans became effective, the English taste for cotton cloth had become well established, and the potential for the mixing of cotton yarns with yarn spun from flax and other fibres was fully developed. Between the Restoration and the mid-eighteenth century, a successful process of import substitution had led to the protected development of the spinning, weaving, bleaching, and dyeing of cotton in Britain and, above all, to profitable experiments with mixtures of cotton and other fibres for the manu-facture of new varieties of cloth, especially Lancashire fustians.

Trade with India and the vacillations of Parliament towards competitive threats to the country's traditional base in woollen textiles, as well as to its younger linen and silk industries, meant that the political and economic precon-ditions for the establishment of an indigenous cotton industry appeared in England several decades before any other economy in Europe. By mid-century, when instabilities and inelasticities in the supply of Irish linen yarn threatened the prosperity of fustian producers and other users of Celtic yarns, the production of cotton for the manufacture of a range of cloth for apparel, bedding, furnishing, and medical and other purposes had proceeded much further in England than elsewhere on the continent, where several states had adopted policies that were either too restrictive (French) or too laissez-faire (Dutch) to allow for successful import substitution.[36]

This helps to explain how and why an embryo cotton industry climbed to a plateau of possibilities, from which accelerated ascent based upon technological breakthroughs seemed likely. Nevertheless, the timing of inventions, their development to the point of commercial viability and their subsequent diffusion within and across industries in order to realize the gains in productivity that flowed from learning by doing and from their adaptation for particular fibres and purposes continue to elude explanation.

---

[35] B.L. Solow, "Introduction", in B.L. Solow (ed.), *Slavery and the Rise of the Atlantic System* (Cambridge: Cambridge University Press, 1991), pp. 1–20.

[36] O'Brien, Griffiths, and Hunt, "Political components", pp. 395–418.

# TECHNICAL CHANGE:
# A PROSOPOGRAPHICAL APPROACH

Our own attempt to repair this deficiency is based upon an investigation into the collectivity of inventors, improvers, craftsmen, and promoters of technological innovation in the industry from 1688 to 1851. Did they perhaps constitute a distinctive body of people who might be differentiated by birth, education, religion, scientific orientation, or entrepreneurial acumen from the population at large? If that turns out to be the case, the inquiry might then move forward to investigate the question why the British Isles, rather than some other European society or culture, had accumulated the type of human capital required to generate such an impressive flow of innovations over that period. Alas, it proved impossible to construct a comprehensive prosopography of inventors, large and small, whose inventions and improvements transformed the productivity of textile production in Britain in this period. Except for a famous minority included in the 2500 or so names culled from patent specifications and other sources, biographical details were hard to compile systematically. Hence our data base can only generate small, random, and possibly unrepresentative samples of innovators to confront the largely unsubstantiated suggestions about the characteristics of British inventors for this period. Nevertheless, our attempt to construct a large sample and to abide by proper statistical procedures might perhaps carry the discussion beyond the ad hoc citation of that familiar list of major inventors who have made their way into textbook accounts of technological change during the industrial revolution.

This is not the place in which to tabulate our evidence or even to cite representative figures. We are inclined to suggest, however, that the composition and attributes of inventors in our sample changed through time. We intend to distinguish innovators active during the period of macro invention and discovery from those who were employed by and large within the textile industry during the decades of improvement and diffusion. In our perception, the beginnings of the latter period coincided with the long wars against France from 1793 to 1815. No precise turning points can be marked in histories of the technologies we considered, but in several potentially illuminating respects, the phase of more haphazard and uncertain discovery (1733 to 1785) seems to differ from the half century down to, say, 1851, when the course of technical change in textiles appears to be more predictable and explicable in economic terms. Historiographically, it is the characteristic features of innovation and of innovators in the latter phase (when almost all branches of textile manufacturing were clearly undergoing or about to undergo technological transformation) that have informed textbook perceptions of the forces and people behind the long-term success of this major industry. But to understand that success story fully, historians need a lens that will both widen and deepen their perspectives of the context for tech-nological progress in textiles. As we have argued, that context is political and cultural as well as economic and must include the factors and people who carried the industry to the level it had attained before the turn of the century. Although difficult to discern and explain, the subset of innovators in this earlier period seems more Schumpeterian and interesting to contemplate than their nineteenth-century successors, who adapted their ideas in response to familiar market forces and pressures.

Our argument, then, is that, in contrast to the "developers" and "adapters" of the later period, the stated aims of the "inventors" and "discoverers" of the eighteenth century tend to be pitched at a level of generality and optimism that suggests a rather widespread, somewhat "pre-professional" interest in new techniques. These men emanated, relative to their successors, in far larger proportions from occupations and locations only tangentially related to the textile industries and to the districts they suggested would benefit from their original ideas.

Although there were exceptions, the collectivity of people involved in the search for new and potentially profitable knowledge tended to come from mainstream class or status categories in eighteenth-century British society. Despite some frequently cited examples to the contrary, the group should no longer be portrayed as dominated by "artisans" from low down the social scale. This popular perception, if it applies at all, relates to the later period of adaptation and development.[37]

It is also clear that the majority of innovators adhered to the Anglican religion. Dissenters figure prominently in the literature, particularly in Weberian-based accounts purporting to uncover religious sources for the development of science and technology in eighteenth-century Britain. But for much of the eighteenth century these so-called "outsiders" are not represented out of proportion to their rather small and declining numbers among the kingdom's literate population. Their names tend to appear in midland textile regions, because counties in that region included a large concentration of nonconformists within their urban communities. In addition, there also seems to be very little that was utilitarian either in the religious beliefs and attitudes of eighteenth-century nonconformists or in the upbringing and education they purchased for their children.[38] Anglican theology and sermons display no antipathy towards innovation, hard work, and

---

[37] This conclusion stems from a prosopographical analysis of textile inventors in the period 1660–1850 available from the Director of the Institute of Historical Research, Senate House, Malet Street, London WC1E 7HU. The data for the occupations of patentees come from B. Woodcroft, *Titles of Patents of Invention, Chronologically Arranged from 1617 to 1852* (2 vols., London, 1854); analysed into status categories using the classifications of K. Honeyman, *Origins of Enterprise. Business Leadership in the Industrial Revolution* (Manchester: Manchester University Press, 1982), and F. Crouzet, *The First Industrialists. The Problem of Origins* (Cambridge: Cambridge University Press, 1985).

[38] M.R. Watts, *The Dissenters. I. From the Reformation to the French Revolution* (Oxford: Clarendon Press, 1978), pp. 287–9, 350–3; E.D. Jebb, *Nonconformity and Social and Economic Life 1660–1800* (London, 1935), pp. 45, 57, 92–3, 112–32; A.D. Gilbert, *Religion and Society in Industrial England. Church, Chapel and Social Change, 1740–1914* (London: Longmans, 1976), pp. 14–108.

business, while the curricula of nonconformist and dissenting academies offered only limited vocational instruction and a modicum of education in natural philosophy and the useful arts. As adolescents, hardly any of our sub-sample of innovators for the eighteenth century received an education that looks, *prima facie*, in any way directly relevant to their subsequent endeavours at the frontiers of industrial technology.[39]

Only a handful of inventors attended universities in England or Scotland. Five or six, however, had participated in local and international networks for the study and exchange of scientific information related to the bleaching and dyeing of cloth. That connexion between science and formal education on the one hand, and the finishing of cloth on the other, is well documented. The flow of British innov-ations for the bleaching and dyeing of cloth which came on stream very late in the eighteenth century depended heavily upon imported knowledge and technicians from Greece, Holland, and France.[40]

Historians have also traced the development of numerous informal and non-institutional channels of communication and education over the eighteenth century. Books, magazines, pamphlets, newspapers, lectures, and exhibitions of toys, models, and automata connected to natural philosophy (science) and technology (mechanics) proliferated.[41]

Scientific societies in which matters of concern to industry were discussed

---

[39] H. McLachlan, *English Education under the Test Acts. The History of the Non-conformist Academies 1662 1820* (London, 1931), pp. 6–15, 43–4; R.S. Mortimer, "Quaker education", *Journal of the Friends' Historical Society* (1947), **39**: 66–70; A.G. Matthews and G.F. Nuttall, "The literary interests of non-conformists in the 18th century", *Transactions of the Congregationalists' Historical Society* (1933–5), **12**: 337–8; A.P.F. Sell, "Philosophy in the eighteenth-century dissenting academies of England and Wales", *History of Universities* (1992), **11**: 75–122. For Anglicanism and innovation, see M.C. Jacob, *The Newtonians and the English Revolution, 1689–1720* (Hassocks: Harvester Press, 1976), and P. Harrison, *"Religion" and the Religions in the English Enlightenment* (Cambridge: Cambridge University Press, 1990).

[40] On bleaching, see A.E. Musson and E. Robinson, *Science and Technology in the Industrial Revolution* (Manchester: Manchester University Press, 1969), pp. 274–331; S.H. Higgins, *A History of Bleaching* (London: Longmans, 1924) pp. 73–9. On Turkey red dye, see H. Wescher, "Turkey red dyeing", *Ciba Review* (December 1959), **12**, no. 135: 21–6; G. Schaefer, "The history of Turkey red dyeing", *Ciba Review* (May 1941), no. 39: 1407–16; *Journals of the House of Commons* **41**: 289, 467, 882.

[41] P. Clark, *Sociability and Urbanity. Clubs and Societies in the Eighteenth Century City* (Leicester: Victorian Studies Centre, University of Leicester, 1986); W.H.G. Armytage, "Education and innovative ferment in England, 1588–1805", in C. Anderson and M.J. Bowman (eds.), *Education and Economic Development* (London: Cass, 1965), pp. 376–93; R.S. Porter, "Science, provincial culture and public opinion in Enlightened England", *British Journal for Eighteenth Century Studies* (1980), **3**: 20–46.

appeared in numerous provincial towns.[42] In the vanguard since 1662, the metropolis had possessed a Royal Society for the promotion of science, but its impractical preoccupations led to the foundation of the Society for the Encouragement of Arts, Manufactures and Commerce in 1754. Within two decades, this metropolitan and gentlemanly body became more active in the promotion of industrial design and the development of technologies relevant for the long-term growth of textiles than the Royal Society, more active, in fact, than any other provincial society, including the Manchester Literary and Philosophical Society, where discussions of practical import were confined to the bleaching and dyeing of cloth.[43] The obvious concerns of the Society of Arts with import substitution, design, and quality and with jobs for underemployed women and children complemented and mirrored the policies of two government bodies, the linen boards of Scotland and Ireland. Conditioned by its own interest in maintaining law and order in rebellious Celtic provinces, the Hanoverian state allocated sums of money to the cultivation of flax, to industrial training, to the diffusion of Dutch techniques for linen bleaching, and even to the provision of spinning wheels for its impoverished Scottish and Irish subjects.[44]

---

[42] See the essays in I. Inkster (ed.), *The Steam Intellect Societies. Essays on Culture, Education and Industry circa 1820–1914* (Nottingham: Department of Adult Education, University of Nottingham, 1985); also I. Inkster, "The provincial context of industrial revolution: science and society in Derby, 1730–1830" (unpublished paper); I. Inkster, "The development of a scientific community in Sheffield 1790–1850: a network of people and interests", *Transactions of the Hunter Archaeological Society* (1973), **10**: 99–131; S.A. Shapin, "The Pottery Philosophical Society, 1819–1835: an examination of the cultural uses of provincial science", *Science Studies* (1972), **2**: 311–36.

[43] D.G.C. Allan, "The Society of Arts and Government, 1754–1800: public encouragement of the arts, manufactures and commerce in eighteenth century England", *Eighteenth Century Studies* (1974), **7**: 434–52; Anon, *A Concise Account of the Rise, Progress and Present State of the Society for the Encouragement of Arts, Manufactures and Commerce, Instituted at London, anno MDCCLIV by a member of the Society* (London, 1763); T. Thomson, *History of the Royal Society from its Institution to the End of the Eighteenth Century* (London, 1812); M. Hunter, *Science and Society in Restoration England* (Cambridge: Cambridge University Press, 1981); and M. Hunter, *Establishing the New Science. The Experience of the Early Royal Society* (Woodbridge: Boydell, 1989); R.A. Smith, *A Centenary of Science in Manchester (for the Hundredth Year of the Literary and Philosophical Society of Manchester)* (London, 1883) pp. 83–5; A.W. Thackray, "Natural knowledge in cultural context: the Manchester model", *American Historical Review* (1974), **89**: 672–709.

[44] Gill, *The Rise of the Irish Linen Industry*; H.D. Gribbon, "The Irish Linen Board, 1711–1828", in L.M. Cullen and T.C. Smout (eds.), *Comparative Aspects of Scottish and Irish Economic and Social History, 1600–1900* (Edinburgh: John Donald, 1977), pp. 77–87; A.J. Durie, *The Scottish Linen Industry in the Eighteenth Century* (Edinburgh: John Donald, 1979); Anon, *Plan by the Commissioners and Trustees for Improving Fisheries and Manufactures in Scotland for the Application of their Funds* (Edinburgh, 1727).

# CONCLUSIONS

Many of the prototype mechanical innovations, preparatory techniques, and finishing processes that eventually transformed all stages and all types of textile production in Britain appeared over the short span of six or seven decades that encompassed the careers of John Kay and Edmund Cartwright, the latter of whom withdrew from textile innovation after 1790. The diffusion, adaptation, and improvement of these eighteenth-century technologies during the reigns of Victoria and her predecessors are not difficult to explain in the terms of demand-led and supply-induced models drawn from economic theory. These models are certainly relevant to the earlier decades of technical breakthroughs simply because (as economists will insist) Europeans at most times were interested in profiting from potentially exploitable discoveries. However, for historians concerned with the interplay of theory, concept, and vocabularies, on the one hand, and specific techniques, machines, artefacts, and varieties of cloth, on the other, demand and supply models have little to offer.

In our view, what has to be explained is the marked rise in the number and range of inventions that emerged in the earlier period. That discernible discontinuity seems to have coincided, moreover, with the development of more pervasive and widespread interest in natural philosophy, mechanics, automata, and even in technological fantasies, among the upper and middle ranks of British society, including members of the ruling élite.[45] Even if these manifestations of a changing culture cannot be measured, they excited comment at the time. Samuel Johnson noticed that "the age is running after improvement. All the business of the world is to be done in a new way".[46] Over these decades, the signs are that English culture was being "reordered". A generalized optimism and spirit of enquiry added to the flow of useful and potentially exploitable knowledge. That same reordering of culture probably raised the propensity of otherwise conservative businessmen to search for and experiment with new ideas and to re-evaluate risks. From this "culture" there emerged a cluster of innovations that eventually transformed the entire production of textiles.[47]

Our preliminary conclusions are that historians who wish to generalize about the sources of innovation for the paradigm case of British textiles are advised to take a long view. They must first explain why the manufacture of all kinds of textiles in general and cotton yarn and cloth in particular had arrived, by the second quarter of the eighteenth century, at an advanced stage of development

---

[45] G. Basalla, *The Evolution of Technology* (Cambridge: Cambridge University Press, 1988), pp. 74–7, for technological fantasies; and P.G. Boucé, "Aspects of sexual tolerance in eighteenth-century England", *British Journal for Eighteenth Century Studies* (1980), **3**: 173–92.

[46] Ashton, *Industrial Revolution*, p. 11.

[47] For cultural historical theory, see M. Thompson, R. Ellis, and A. Wildavsky, *Cultural Theory* (Boulder, Colo., and Oxford: Westview, 1990).

containing the seeds of further progress. That explanation will surely be dominated by an analysis of the political economy of overseas trade with Asia, Africa, and the Americas. There may be space in this phase of technological history for a distinctly English culture of consumerism or a fashion-driven propensity of the urban middles classes to buy more clothing and fabrics. "Supply side forces" which integrated the domestic economy and provided households with the real incomes required to spend more on textiles, however, still seem to be altogether more important as factors behind the extension of the home market for clothing. The development of foreign markets and foreign imports of raw materials for domestic textile production continued to depend upon a framework of mercantilist regulations enacted by aristocratic governments in London. Finally, when they turn to study actual inventors and inventions one by one, unless they wish to revert to the heroic and theological vocabularies of the Greeks and Victorians, historians of technological change in textiles will be moved inexorably to consider that portmanteau category of culture and to reach for Josiah Tucker's observation that a "strange frenzy has infected the whole English nation".[48]

The harmonious integration of science and religion into "culture" was expressed and exemplified in verse by a latitudinarian Anglican clergyman, the Reverend Edmund Cartwright, inventor of the wool comber and power loom, macro inventions by any standards.

> Since even Newton owns that all he wrought
> Was due to industry and patient thought
> What shall restrain the impulse that I feel
> To forward as I may the public weal
> By his example fired to break away
> In quest of truth through darkness into day.[49]

---

[48] There were calls for Lancashire poets to compose an epic of invention or an "Arkwrightiad". One appears in F. Espinasse, "Lancashire industrialism: (A lecture to the Mechanics Institution, Manchester)", *The Roscoe Magazine* (1849), **1**: 201–9, on pp. 206–7.

[49] M.S. [M. Strickland], *A Memoir of the Life and Writings and Mechanical Inventions of Edmund Cartwright, D.D., F.R.S., Inventor of the Power Loom etc.* (London, 1843), p. 279.

**Technology, Politics, and National Cultures**

# Technology Transfer and Industrial Transformation: An Interpretation of the Pattern of Economic Development Circa 1870–1914

*Ian Inkster*

## ON THAT WHICH NORMALLY FAILS

Technology transfer is most often thought of as a very modern process. Even when our focus is firmly centred on the transfer of techniques between nation states (as against its transfer between industries or sectors within one economic system), attention is commonly drawn towards the relatively recent accelerations of growth that have seemingly arisen out of the transfer of technique from the USA to Europe and Japan. Just as recent are the more problematic, less successful transfers of technology from developed economies to those poorer nations attempting a programme of speedy economic modernization.[1] Somewhere between these extremes lies the recent experience of the newly industrializing

---

[1] Of course, failure is less commonly studied. For recent work, see David J. Jeremy (ed.), *International Technology Transfer. Europe, Japan and the USA, 1700–1914* (Aldershot: Elgar, 1991); David J. Jeremy, *The Transfer of International Technology. Europe, Japan and the USA in the Twentieth Century* (Aldershot: Elgar, 1992); D. Charles and J. Howells, *Technology Transfer in Europe. Public and Private Networks* (London: Belhaven, 1992). The term is often used to examine problems arising from the movement of technique between public sectors and industry, or between military and civilian functions; for a study of this in the case of China, see Detlef Rehn, *Shanghais Wirtschaft im Wandel, mit Spitzentechnologien ins 21. Jahrhundert* (Hamburg: Institut für Asienkunde, 1990).

countries, especially the so-called "Asian tigers" — South Korea, Singapore, Hong Kong, and Taiwan — whose programmes of technology transfer and export promotion have been associated with the fastest rates of overall economic growth in the world.[2]

Yet technology transfer normally fails, and this is as true of recent times as it is of the distant past. It is, indeed, for precisely this reason that we may postulate that the combination of technological change and technology transfer has been a major determinant of the pattern of world development and underdevelopment from the eighteenth century to the present.[3] The very particularity and peculiarity of both technological genesis and its creative adoption elsewhere has helped to forge a world of diversity, of tragedy and triumph, of poverty and prosperity. The movement of technologies has, clearly, never been osmotic. Capital and the capturing of intellectual property rights have never represented a sufficient combination for the successful transfer of either specific techniques or technological systems from one nation to another. This is so, even in cases where transfers have been tried between nations whose economic and institutional systems have had much in common. In post-war Europe and Japan, transfer *in* from the USA was only possible in association with tremendous institutional and attitudinal responses and adjustments.[4] Indigenous technological systems, consisting as much of institutions, regulations, and cognitions as of mechanical or other techniques and artefacts, are evidently complex and distinctive,

---

[2] A. Hernadi, "Export-oriented industrialisation and its success in the Asia-Pacific region", *East Asia* (1985), **3**: 76–93; R. Wade, *Governing the Market. Economic Theory and the Role of Government in East Asian Industrialisation* (Princeton, NJ: Princeton University Press, 1990); Robert Elegant, *Pacific Destiny. Inside Asia Today* (New York: Crown Publishers, 1990); Nigel Holloway (ed.), *Japan in Asia. The Economic Impact on the Region* (Hong Kong: Review Publishing Co., 1991); S.B. Schlossstein, *Asia's New Little Dragons* (Chicago: Contemporary Books, 1991); World Bank, *The East Asian Miracle. Economic Growth and Public Policy* (New York: Oxford University Press, for the World Bank, Washington D.C., 1993). For a brilliant and accessible sweep, see Ezra F. Vogel, *The Four Little Dragons. The Spread of Industrialization in East Asia* (Cambridge, Mass.: Harvard University Press, 1991).

[3] See Ian Inkster, "Mental capital: transfers of knowledge and technique in eighteenth century Europe", *Journal of European Economic History* (1990), **19**: 403–441; Ian Inkster, *Science and Technology in History. An Approach to Industrialisation* (London: Macmillan, 1991).

[4] E.A. Brett, *The World Economy Since the War* (London: Macmillan, 1985); E.F. Denison, *Why Growth Rates Differ* (Washington: Brookings Institution, 1967); Moses Abramovitz, "Catching up, forging ahead and falling behind", *Journal of Economic History* (1986), **46**: 379–96.

highly unlikely at any time to match the technical requisites of imported techniques.[5]

Despite the fact that policy makers must view historical experience in technology transfer as poor evidence of its rewards, it is precisely this tactic that appears at the heart of plans for radical economic modernization. From Independence onwards, Indian policy makers placed tremendous faith in the transforming power of technology transfer, just as they recognized the need for profound changes in institutional structure. As the First Five Year Plan put it, certain forms of indigenous "economic and social organisations are unsuited to or incapable of absorbing new techniques and utilising them to the best advantage".[6] One of the findings of the famous Mahalanobis Committee of the early 1960s was that existing large Indian enterprises were gaining special access to foreign technology, to the detriment of smaller firms and the workshop economy, and that a major tactic should be to forge complementarities between advanced foreign technologies and rural and smaller-scale technique.[7] Belief in the power of foreign technology remained but was now formulated with an eye to the very recent experience of the only Asian "miracle economy" of the 1960s:

> ... import of know-how, particularly process know-how or product design, should continue to be allowed on a discriminatory basis, so that Indian industry is able to keep in touch with the world technological mainstream ... it should be the endeavour of Indian research and development to build and develop it to suite Indian conditions, Indian environment and Indian raw material availability. This in fact is what Japan has done with great advantage to her economy.[8]

---

[5] For formal notions of technological system, see W.E. Bijker, T.P. Hughes, and T.J. Pinch (eds.), *The Social Construction of Technological Systems. New Directions in the Sociology and History of Technology* (Cambridge, Mass.: MIT Press, 1987). For excellent detailed examples of how the introduction of even quite discrete techniques into systems similar to those of the technology of origin nevertheless required considerable micro-adaptations, cumulative and painstaking, see J.A. McGraw, *Most Wonderful Machine. Mechanisation and Social Change in Berkshire Paper Making 1801–1885* (Princeton, NJ: Princeton University Press, 1987); K. Bruland, *British Technology and European Industrialisation. The Norwegian Textile Industry in the Mid-Nineteenth Century* (Cambridge: Cambridge University Press, 1989).

[6] Bepin Behari, *Economic Growth and Technological Change in India*, (Delhi: Delhi Press, 1974), quotation on p. 132.

[7] Beldav Raj Nayar, *India's Quest for Technological Independence. Policy Foundation and Policy Change* (Delhi: National Publishing House, 1987); B.D. Nag Chaudhuri, *Technology and Society. An Indian View* (Simla: Indian Institute of Advanced Studies, 1979).

[8] *Report of the Committee on Foreign Collaboration. May 1967* (New Delhi: Ministry of Industrial Development and Company Affairs, 1968), quotation on pp. 2–3.

In summary, experience suggests that technology transfer is a potentially powerful mechanism of industrial revolution, that technology transfer normally fails, and that the creative effort and resource mobilization involved in successful technology transfer is in most senses quite equivalent to processes associated with initial, significant technological change within the core of advanced systems. At the same time, it is known that processes of technology transfer which weaken and swamp existing indigenous technological systems have been, in the long term, disastrous, and that this lesson is to be learnt from experience both recent and antique, both Asian and European.[9]

## WINNERS AND LOSERS IN THE FIRST CLIMACTERIC

The years after the Crimean war, and especially after the Franco-Prussian war were ones of international industrial and commercial competition, imperial rivalry, and technological transfer, as statesmen and bureaucrats moved along a learning curve that described a very strong and hardening relationship between military success and industrial capacity. If around 1880 total industrial production in the new Germany was approximately one third of that of Britain, by 1913 that nation had overtaken the leader. Although in per capita industrial production Britain remained ahead, in certain areas of fairly sophisticated technique Europe and North America surged forward. The USA moved from approximate parity with Britain in steel production around the mid-1880s to a position in 1912 when annual steel production in the USA was running at 4.5 times that of Britain. Again, the dramatic rise of the German heavy chemical industry, revolving around a rapid application of the new Solvay process of soda manufacture, was associated not only with technology transfers but also with novel German technological innovations and organizational formats, the formation of enterprise-based research groups engaged in aggressive patenting and marketing, and the subsequent emergence of the electro-chemical industry.[10] The mantle of technological modernity seemed to have passed quite rapidly from Britain to the USA and Germany and a small coterie of "winner" economies in which Austria, Hungary, Italy, Sweden, Russia, and Japan figured most prominently.

This industrial transformation of global proportions (global in terms of its range and its commercial, technological, and political impacts on most non-

---

[9] Nathan Rosenberg, *Inside the Black Box. Technology and Economics* (Cambridge: Cambridge University Press, 1982), chapter 11; Ian Inkster, "Prometheus bound: science, technology and industrialisation in Japan, China and India — a political economy approach", *Annals of Science* (1988), **45**: 399–426.

[10] For basic data, see P. Bairoch, "International industrialisation levels from 1750 to 1980", *Journal of European Economic History* (1982), **2**: 280–96; L.F. Haber, *The Chemical Industry During the Nineteenth Century* (Oxford: Oxford University Press, 1958); W.W. Rostow, *The World Economy. History & Prospect* (Austin, Texas: University of Texas Press, 1978); Inkster, *Science and Technology in History*, pp. 110–23, 131–9.

industrializing nations and regions) might be termed the first of the modern economic climacterics, a massively critical period in the natural history of the world-wide process of industrialization. While this climacteric may or may not have been characterized by the loss of strength within the industrial leaders, it is far less questionable that a major medium of transferred development was the movement of technology across the entire range of production, from agricultural machinery, through industrial prime movers, to entire transport systems (especially railways).[11] It cannot be our purpose here to provide an explanation or strong interpretation of the locus of change, the mechanism of its spread, or the reasons for its stark limitation prior to 1914. Instead, we may more properly address the central importance of technology transfer in the climacteric and demonstrate certain features associated with two major cases of industrialization through transfer, those of Japan and Russia.

## CASE STUDIES FROM JAPAN AND RUSSIA

Whatever the previous level of economic advancement in the Tokugawa era, as Kazushi Ohkawa has summarized, "Japanese modernisation — economic, political and social — began, at least symbolically in 1868 when the Emperor Meiji was restored to the throne".[12] It seems very clear that the first years of the Meiji period (to around 1881–4) were ones of enormous institutional innovation, characterized by a series of changes designed to remove feudal privileges without creating populist revolution, to formalize and centralize the collection of govern-ment revenues, and to establish new property rights. Finally, it may be judged that such institutional innovation was in some complex way requisite to the speedy introduction of a range of infrastructural, military, and industrial techniques from the West.[13] In the years that followed, industrialization continued under the auspices of a massive governmental military programme and the continued absorption of surplus labour from agriculture into the industrial and services sectors of the economy.[14]

The Russian case is clearly different. Although the spurt of industrial growth in the years approximately 1887–1913 is unquestioned (per capita industrial output doubled, as did foreign trade and employment in mining, railway, and

---

[11] For a study of Britain as the centre of a new system of technology transfer, see Rosenberg, *Inside the Black Box*, chapter 11.

[12] Kazushi Ohkawa, "Capital formation in Japan", in Peter Mathias and M.M. Postan (eds.), *The Cambridge Economic History of Europe. Vol. VII: The Industrial Economies, Part 2* (Cambridge: Cambridge University Press, 1978), pp. 134–65, on page 142.

[13] Ian Inkster, *Japan as a Development Model?* (Bochum: Studienverlag Brockmeyer, 1980), especially chapters 2 and 3.

[14] E.H. Norman, *Japan's Emergence as a Modern State* (New York: Institute of Pacific Relations, 1940); Ryoshin Minami, *The Economic Development of Japan. A Quantitative Study* (London: Macmillan, 1986).

manufacturing), the tremendous difficulties of industrial diffusion within the setting of such a huge and hard terrain, compounded with the institutional backwardness of the agricultural sector, meant that the overall performance of the economy was mixed. Industrialization was enclavist.[15] In Russia, a very major component of industrial modernization was the state-instigated railway, mining, and metallurgical and metalworking project in South Russia, composed of a great series of technological transfers involving European and American labour and engineering skills, entrepreneurs, financing and firms, joint ventures, and government guarantees, all primarily located within an area on the north coast of the Black Sea bounded by Odessa, Kharkov, and Rostov. This was the advanced technological site which best represented Finance Minister Sergei Witte's vision for his nation and which best witnessed the playing out of Podkolzin's "medley of various modes of production".[16] Whereas in Japan technology transfer became associated with a highly generalized process of economic transformation, in Russia a more concentrated effort led to a dualistic economic structure and a loss of political control at the turn of the century which was worsened by the war with Japan (1904–5), followed by a short but more broadly based industrial growth in the years 1908–13, a period of relative success brought to a close by the twin forces of war and revolution. The material on Japan and Russia that follows does not purport to provide a general survey of technology transfer to the two nations, much of which has already been delineated.[17] Rather, the task here is to uncover salient features of the transfer process which illustrate the key moments of success and the importance of specific choices, mechanisms, and sequences.

## THE SHIP OF STATE: JAPAN

Under the heading of statist scurrying might be demonstrated the vital importance of the Japanese government's non-commercial and multifuctional inter-

---

[15] R.W. Goldsmith, "Economic growth of Tsarist Russia 1860–1913", *Economic Development and Cultural Change* (1961), **9**: 441–75; Paul Gregory, "Economic growth and structural change in Tsarist Russia", *Soviet Studies* (1972), **23**: 418–34; Olga Crisp, *Studies in the Russian Economy before 1914* (London: Longman, 1973).

[16] A. Podkolzin, *A Short Economic History of the USSR* (Moscow: Progress Publishers, 1968), quotation on page 45.

[17] In addition to references under my first section, above, see the special issue of *The Developing Economies* (1977), **15**, no. 4, devoted to "Adaptation and transformation of western institutions in Meiji Japan"; A.W. Burks (ed.), *The Modernisers. Overseas Students, Foreign Employees and Meiji Japan* (London: Wiley, 1985); C.E. Black (ed.), *The Transformation of Russian Society* (Cambridge, Mass.: MIT Press, 1960), and Ian Inkster, *Science and Technology in History*, pp. 42–66 and chapter 7.

ventions in the key years from 1868 to 1885. These ranged from original "search and buy" forays to the laying down of physical infrastructure incorporating Western techniques (docks, harbours, lighthouses, railways), training and skilling provisions, the direct employment of foreign experts at many levels, and the establishment of a system of social and political control which dampened effective resistance to new technologies. Between 1868 and 1895 some 2,500 students and officials were sent by the government to Europe and North America to study and investigate best techniques, and prior to 1910 Japanese technicians, artisans, and merchants participated in at least 38 major international exhibitions with the financial assistance of the government.[18] The government's own model factories were not merely designed as demonstrations but as part of the search and screening process. Established in 1874, its Akabane Engineering Office was designed to construct selected and tested machinery for purchase by private enterprise, a function also performed by the government printing office.[19]

Particularly in areas central to state interests (transport, shipping, and the military) the government set up a complex intelligence network in order to search out expertise, and a series of internal filters designed to exploit foreign employees to the advantage of industrial modernization. Once searched and screened, foreign employees were not mere technical go-betweens (itself an essential role); they often emerged as vital elements in a continuous learning curve. Western science professors at the Imperial College of Engineering, for example, stuck to neither science nor the classroom. From an early stage, their Japanese students were closely involved in projects for mineral exploitation, surveying, medical and sanitary reform, national seismological research, mechanical engineering, architecture, and so on. E. Piquet, a non-government silk inspector of the firm of Weber, Hall and Co., was highly influential in determining government action in controlling the quality and volume of export of silk cards and raw silk, and publicized his findings through such associations as the Silk Committee of the Yokohama Chamber of Commerce. Dr P. Mayet, although employed at the Education Department, so impressed the bureaucrats with his lengthy scheme for insurance in Japan that he was appointed to the Home Department in order to explore possible insurance reform. From his new base, Mayet went on to propose changes in Japan's financial policy, gave detailed arguments in favour of the level of internal debt, and was rewarded through promotions to the position of adviser to the Financial Section of the newly-established Council of State. Here the foreign influence was at the very centre of power.[20]

---

[18] For these and other instances see Inkster, *Science and Technology in History*, pp. 187–93.

[19] Similar functions seem to have been performed by the Japanese Mint also.

[20] *Shyrio Oyatoi Gaikoku-jin* (Tokyo: Shogakukan, 1975): 420–2; *Oyatoi Gaikoku-Jin Seimi Kyoryo Kigen, Shokumu-Ichiran* in Yoshino Sakuzö, *Meiji Bunka Zenshu*, vol. 16 (Tokyo: Nihon hyöron-sha, 1928), pp. 347–62.

In partnership with private enterprise, the Japanese government utilized the detailed catalogues of British, American, or German machine works to derive exact specifications, and all purchases were on the competitive tender principle. In the early years, most machinery imports were arranged or handled by foreign merchants, many operating from the open ports of Japan. Table 1 gives some idea of the relatively small number of firms concerned in the early years, and of the dominance of Britain.[21]

Such firms represented information systems, potentially available, at a price, to Japanese agents in both the public and private sectors. By the mid-1890s, British firms were losing out to American and then German machine suppliers, as the Japanese tracked the best choices by way of the foreign merchants. Japanese government inspectors were everywhere, reportedly working under the motto "if in doubt, reject". As one British machine supplier complained of imported equipment, "It is no unusual thing for them to be pulled apart entirely and for the whole of the paint to be scraped off the surface of the castings and other parts."[22] At the same time, well trained if yet inexperienced Japanese advisers based in London regularly sent out designs and specifications of best

**Table 1:**   Foreign Firms in the Five Open Ports of Japan (1874–1883)

|      | British | American | German | French | All Other Western | Chinese | Total |
|------|---------|----------|--------|--------|-------------------|---------|-------|
| 1874 | 155     | —        | —      | —      | 215               | 95      | 465   |
| 1875 | 109     | —        | —      | —      | 148               | 84      | 341   |
| 1876 | 80      | —        | —      | —      | 141               | 72      | 293   |
| 1877 | 83      | —        | —      | —      | 149               | 53      | 285   |
| 1878 | 92      | —        | —      | —      | 151               | 40      | 287   |
| 1879 | 90      | —        | —      | —      | 141               | 89      | 320   |
| 1880 | 108     | —        | —      | —      | 150               | 102     | 360   |
| 1881 | 109     | 49       | 36     | 42     | 32                | 78      | 346   |
| 1882 | 98      | 39       | 37     | 17     | 17                | 252     | 460   |
| 1883 | 91      | 94       | 34     | 16     | 24                | 346     | 545   |

---

[21] The table is constructed from various data, often contradictory, found in Japanese trade directories, in particular *Japan Herald Directory and Hong List* (Yokohama, 1870); *Japan Daily Herald Directory and Hong List* (Tokyo, 1872); *The Tokyo and Yokohama Directory for the Year 1881* (Yokohama, 1882); *The Tokyo and Yokohama Directory* (Tokyo, 1889).

[22] Screening of technological imports took place in the context of a significantly increasing bill for imports. Machinery imports rose from one seventh of total manufactured imports in 1886 to one third in 1890 and one half in 1895. By the end of Meiji, annual average imports of machinery, parts, and instruments had reached nearly 40 million yen.

technique to almost all the major Japanese merchant houses, especially in the field of heavy equipment designed for government armaments and transport systems. The frequent trade commissions sent to Japan from Europe and America obtained only limited results unless they went through the merchant and bureaucratic screening nexus. At the same time, Japanese external search was becoming more sophisticated and less dependent on government intervention; witness the large party of Japanese engineers and metallurgists who toured steelworks in America and Europe during 1896.[23]

A second characteristic of the Japanese case may be captured in the notion of a moving interface. Technology transfer did not occur as an "impact" as such, but rather over a long period within a multi-layered interaction involving a variety of agencies and institutions. Although much attention has been devoted to interactions deliberately fostered by the Japanese government, there is a strong case for an alternative strategy that would place the anarchic and incremental influence of the "creeping frontier" in the treaty ports at centre stage. Only thereby can the effectiveness of bureaucratic mechanisms be understood. Thus, those merchants such as J. Favre Brandt or Ahrens and Co., who were involved in specific introductions of Western equipment, who were fluent in Japanese and well connected to Japanese bureaucracy, obtained their strategic positions and advantages through the complex interactions with Japanese and other foreigners within the institutional frame of the treaty ports.[24] The treaty ports have been seen as emporiums of Westernism, but they were also entrepots for Western attitudes, knowledge, and technique.

Western agency was greatly assisted by a third characteristic of the Japanese case: the extremely high status of transfer agents. The popularly acknowledged position of the Japanese élite as itself a "live learning machine" was well illustrated in the song sung by Tokyo University students in the early 1870s: "He's a student! He's a student, in disparagement they say: Yet the present Privy Council, what but students all are they?"[25] An excellent instance of the elite's symbolic representation of the West was provided in the Great National Industrial Exhibition of 1881, which began and ended with speeches from the Emperor. The distribution of awards for technical or commercial excellence and the major award speech to the 31,000 exhibitors were made by the Emperor Meiji,

---

[23] *Daily Chronicle* (London), 4 March 1896, p. 7.

[24] Brandt had been first attaché to the Swiss trade mission in Japan from 1862. He began supplying arms, munitions and machinery to government and from the early 1870s supplied machinery to Japanese private enterprise and engaged formally in instructing buyers in usage and best practices. Ahrens opened in Japan in 1870 and maintained an important link between German and Japanese officialdom, as its principals were attached to the German consulate service throughout.

[25] Reproduced in *Japan Weekly Mail*, 12 September 1870, pp. 437–8.

surrounded underfoot by princes and nobility, officers of state, and Western diplomats and consuls. A little further afield were the Western residents and Japanese spectators, mixed in with a splash of Korean guests, the bright robes of the latter forming "an agreeable oasis in a desert of black and neutral tints". With daily visitors numbering some 10,000 to 20,000, the exhibition acted as an enormous international industrial mart, illustrating, in the words of Prince Kita Shirakawa, "how sensible an impetus exhibitions impart to manufacturing diligence". The emperor's own main theme was the need to "develop a spirit of industry and perseverance". Prize-winners included ex-nobles and ex-samurai, modern factories and filatures, as well as individual craftsmen in ivory, enamelling, and metal working.[26]

A further notable aspect of the Japanese case lay in the potency of military demands which rose throughout the Meiji years, culminating in the technical needs of the wars with China (1895) and Russia (1904–5). Within the emerging military-industrial complex, new technologies transferred especially quickly: examples are the water jacket blast furnace (USA 1882–Japan 1890), the first successful application of the Bessemer steel process to copper refining (USA 1884–Japan 1893), and electrolytic refining (USA late 1880s–Japan mid-1890s). Transfer-diffusion effects accelerated after the demonstrated success against China during 1895. Henceforth, Japan was to receive foreign capital at will. By 1902, arsenals held over 80 per cent of total horsepower capacity in the Japanese machinery industry and employed half the sector's workers. Furthermore, the military sector was by far the most westernized and served to directly stimulate the emergence of larger-scale Japanese metallurgy and metalworking enterprises during the early twentieth century.[27]

The last major clear characteristic of the Japanese case, often confused in the literature on Meiji industrial development, is that of private sector followership in the overall process of technology transfer. The large-scale or significant parti-cipation of the private sector in the financing and instigation of technology transfers awaited the later 1880s for a variety of reasons. Among such reasons, however, were the risk-securing and risk-reducing impacts of renewed social control measures after the disarray of the Satsuma rebellion (1877), the demon-stration effects of model factories and arsenals, the cheapening of technology and capital resulting from the government's selling-up process during 1881–5, an increased supply of skilled labour emerging from earlier projects and training schools, and the cost-reducing impacts of state-run infrastructural services, especially transport and haulage. This is, of course, a somewhat tedious way of claiming that state interventions in the field of technology transfer and settle-

---

[26] *Japan Weekly Mail*, 15 August 1881, pp. 386–9.

[27] Nihon Sangyo Kikai Kogyo-kai, *Sangyo Kikai Kogyo hatten Katei* [Development Process of Industrial Machinery Industry] Tokyo: NSKK, 1965); R.H. Smith, "Engineering and education of engineers in Japan", *Tokyo Daigaku*, 12 July 1877, reproduced in *Japan Weekly Mail*, 14 July 1877, pp. 605–8.

ment created the conditions for private enterprise innovation and technology transfer after the mid-1880s.

## CAPITAL AND TECHNIQUE: RUSSIA

The South Russian project of the years approximately 1885–1901 represented a major shift in the industrial energy of Russia towards the Black Sea. During the late 1890s, the Black Sea ports of Odessa and Nikolayev were capturing a great range of industrial productions and fabrications, flanking outwards from the initial core developments in mining, metallurgy, and machine construction; to iron plates, shipbuilding and engineering products, cannon, steam engines, locomotives, and artillery were added glass ware, foodstuffs, refined sugar, soap and glycerine, and cement. By the turn of the century, over 50 per cent of all foreign capital in Russia was located in this area, and Russia had reached fourth place in the global production of iron, 50 per cent of its output coming from the South Russia project.[28] Yet in 1885 the Russian iron industry was technologically backward and hounded by problems concerning location and the supply of raw materials.

From the last years of the eighteenth century, Russian governments had periodically tried to establish a mining and pig iron industry in the South. Lugansk (north of Rostov) was the site of early attempts in South Russia to produce pig iron from coal and coke, as an alternative to the costly Urals wood-using iron processing techniques.[29] In the 1840s, government-sponsored experiments at Kerch (on the south coast of the Sea of Azov) succeeded in using the anthracite coals of the Groushensky mines. German expertise appears to have had some influence in removing further efforts to Ekaterinoslav with the establishment of the Petrovsk works. Refractory furnace materials caused technical problems, and in 1866 the government renewed efforts at Lugansk to

---

[28] The traditional location of the Russian iron industry was the Urals, established in government works of the seventeenth century which harnessed the skills of German workers. With the encouragement of Peter the Great and the transfer of government mines of Neviansky to private hands in the early 1700s, there was a general spread of ironworks, including the foreign entrepreneurial establishments of Henning at Olonets and Ekaterinburg. Between 1719 and 1782 some 150 works were founded.

[29] Wood was used in Urals metallurgy as well as being a locomotive fuel. Whereas in Europe a blast furnace might be located close to both coal and iron deposits, in Russia a single blast furnace required an area of 20,000 acres of forest for fuel. River transport was also very expensive. After several abortive attempts a Bessemer plant was installed in the Urals in 1875–6 under private auspices using French technicians and machinery with some components from Russian suppliers. For developments before 1861, see T. Esper, "Industrial serfdom and metallurgical technology in 19th century Russia", *Technology and Culture* (1982), **23**: 583–608.

exploit nearby coal deposits.[30] From 1870, it was clear that local coal coked badly, and the distance from the iron ore increased costs further. In 1868 John Hughes was invited to set up a £300,000 blast and rolling plant at Ekaterinoslav in exchange for government privileges — grants of crown land, the construction of the Constantinovka railway, and a ten-year premium on production.[31] From then until the 1880s, the Hughes's Novorossiisk (New Russian) Works competed in South Russia only with the Groushensky works which utilized anthracite and brown ironstone.[32] In 1874 a congress of Russian coal proprietors attempted to find ways of exploiting the metallurgical potential of the Donetz coal beds, and throughout the 1870s the Russian state financed prospecting and research teams, produced a geognostic map, and finally discovered the richness of the anthracite deposits of the Eastern Donetz and the Don Cossacks. In summary, until the mid-1880s the major attempts to modernize metallurgical technology and to remove it from the confines of Urals technique were made by government and they failed primarily on the grounds of high transport costs. By the 1880s, the needs of the state were such as to project a much greater effort.[33] In particular, the scarcity of wood almost certainly triggered the South Russia project: by the mid-1880s, wood fuel prices had risen to such a level that English coal around Moscow was selling at well below the domestic wood fuel prices. In the absence of governmental plantations, the Urals technique had led to a reckless reduction of forest resources.[34]

---

[30] Lugansk received cast iron from the Urals at enormous cost.

[31] Earlier attempts at pig iron production in the Donetz utilized French and German artisans and workers from Siberia and failed. Early concessions failed until Hughes founded his limited liability company in London with British capitalists, which from 1869 used all British equipment including rolling mills for rails and coal pits. Russian government regulations which specified steel on all railways prompted transfers of Siemens-Martin technology and government contracts for 700 tons of rail weekly.

[32] See generally J.P. McKay, *Pioneers for Profit. Foreign Entrepreneurship and Russian Industrialisation 1885–1913* (Chicago and London: Chicago University Press, 1970); W.L. Blackwell, *The Beginnings of Russian Industrialisation 1800–1860* (Princeton, NJ: Princeton University Press, 1968); J.M. Crawford (ed.), *The Industries of Russia* (St Petersburg: Published for the World's Columbian Exposition, Chicago, 1893).

[33] The failure of the 1861 Emancipation to spark fundamental institutional changes in agriculture combined with high import bills for basic raw materials: in 1869 Russia produced around 14.4 million poods of iron but imported an additional 18 million poods. Imports of machinery were also a major concern, costing Russia some 8.8 million roubles during 1857–64, 11.9 million roubles in 1865–8 and 23.8 million roubles in 1869–72. For transport see M.L. Tegoborski, *Commentaries on the Productive Forces of Russia* (London: Longman, 1855); P.I. Lyashchenko, *History of the National Economy of Russia* (New York: Macmillan, 1949).

[34] *The Engineer*, 24 July 1885, p. 63.

Situated on the Dnieper directly north of the Krivoi Rog and west of the Donetz basin, Ekaterinoslav and its region boasted iron ores yielding 70 per cent iron, coking and anthracite coal, manganese and limestone, dolomite, and other metallic ores, and by 1895 was identified by George Kamensky of St Petersburg as the potential world leader in iron-based metallurgy and metal-working industries.[35] Ekaterinoslav emerged as the epicentre of a project that began with a complexity of government intervention. New high duties on foreign pig iron and the government's construction of the Ekaterinsky railway joining the Krivoi Rog (iron) with the Donetz coal basin (involving 265 miles of carriage) instigated a series of further public and private responses in a very short time-span. In 1893 the further extensions by government of rail branches already drew for rail supply from the local works of Kamensky and Briansk.[36] There were two waves of private sector followership, covering approximately the years 1887–94 and 1895–1901. The first phase was led by Briansk and Co., an enterprise founded in 1873 but which opened its South Russian Alexandrovsky ironworks in 1887 with the first successful blast furnaces of the region. This was, at first, a complete turnkey project, incorporating a direct transfer of iron and steel-making capacity from Orel (directly north, equidistant between Smolensk and Kharkov) to the iron and coal resources of the south. By the early 1890s, the enterprise was a highly modernized one, incorporating a large mining element, Coppée ovens for coke production, a full Bessemer plant, rolling mills, and an open hearth for acid and basic steel production, and employing over 3,700 men.[37] Also within this first wave of activity were the Kamensky works of the South Russian Dnieper Metallurgical Co. (1889), the Gdantsevsky works of the Krivoi Rog Co., which began smelting in 1892, and the Drougekovsky works of the Donetz Co., situated nearby to the Hughes establishment, which began operation in 1894.

Within the first phase, all enterprises depended on government encouragement, privileges, and contracts, foreign expertise and joint venture, and large

---

[35] George Kamensky, "The ironworks of the south of Russia", *The Engineer*, 6 September 1895, pp. 241–2.

[36] Major increases of duty had been put on steel rails from the mid-1870s for explicitly military reasons; see *The Engineer*, 14 May 1875, p. 334. For the importance of branch lines, see *The Engineer*, 31 August 1894, p. 193.

[37] George Kamensky, "The ironworks of the south of Russia", *The Engineer*, 20 September 1895, pp. 293–5. Turnkey experiments were being tried elsewhere, as at the Polowzev rail-rolling mill in Perm, far to the east of Moscow; a great complex of technique based on charcoal blast furnacing (open hearth furnaces, Erdmann rolling mills etc.) was supplied by Belgian, French, and German firms, transported over a distance of nearly 5,000 miles; see *Engineering* (1894), **60**: 304–5. Ural iron manufacture was clearly influenced by the opening of rail links to Siberia; see *Stahl und Eisen* (1894), **15**: 485.

imported techniques and technological systems. The New Russian works benefited greatly from government railway and port expansion. Thus the planned completion of the port of Marinpol and the canal connexion planned between the Sea of Azov and the Black Sea began a fever of activity at the Hughes company. This included the sinking of new pits, imports of pumping machinery from English engineering firms, an increase in blast furnace capacity, the augmentation of steel production by the erection of additional Siemens-Martin smelting furnaces and new cogging mills, and the establishment of new rail workshops.[38] By 1894, the New Russia Co. utilized a sophisticated technology combining low-grade local ores with those of the Krivoi Rog and employing 7,500 men.[39] Whereas the Donetz Company's coke-making, Bessemer department, open hearths, rail mills, and engineering shops, employing 2,000 men in total, had been founded as a company in Paris (with an initial 10 million francs raised), the Kamensky Works were established as a joint venture of the Belgian firm of Cockerill and the Warsaw Steel Co. Again, this was initially something of a turnkey operation, with the Warsaw firm transferring its entire plant. The complete range of best technique was once more in evidence by the end of this phase. The Kamensky works included a full Bessemer department, an open hearth department, a steel rolling mill, and plate sheet and iron mills. By 1894, over 3,600 men were employed, excluding those involved in mining and coking.[40] The works of the Krivoi Rog Co. concentrated on smelting low grade ores in order to produce foundry pig iron for sale to the south and central governments of Russia.[41]

The second phase of company expansion was somewhat less dependent on direct government intervention and represented more of a partnership between Russian and foreign interests. The second phase generated a greater range of production, increasingly away from the railway sector towards shipbuilding and general engineering activity, and in essence represented the establishment of a modernized capital goods sector within the Russian empire. Whilst the first group relied almost entirely on the transport and natural resources of the Ekaterinoslav and Krivoi Rog region, the later enterprises were more geographically dispersed or more centred on the Donetz coal fields.

By the end of 1896, new companies included Belgian works at Taganrog, Russian works at Lugansk, a branch of a Belgian blast furnace company

---

[38] *The Engineer*, 11 March 1887, p. 312.

[39] Kamensky, "The ironworks of the south of Russia". Hughes did not utilize the Bessemer process, rails being produced entirely from open-hearth steel. Annual capacity by the early 1890s was 100,000 tons of pig iron, over 50,000 tons of rail steel, and 150,000 tons of coke.

[40] This company owned iron mines at Krivoi Rog, manganese mines, limestone mines in the Donetz and bought in large quantities of dolomite and coke.

[41] Kamensky, "The ironworks of the south of Russia", p. 293.

(Olkhoval Co.), and several wholly Russian companies. As early as 1891, there was a notable reduction in iron and steel imports from Europe, and this was seen at the time as a cause of the contemporary industrial recession in eastern Germany.[42] By 1896, the two groups of companies were together producing some 60,000 tons of pig iron annually. The 23 major companies were operating with an estimated capital stock of some 83 million roubles, and there was a further expansion of capacity between 1897 and 1900; by 1901, South Russia and the Urals produced 4 million tons of iron ore annually, compared with Britain's 14.5 million tons.[43] By that time, well over sixty metallurgical and related companies were at work in the Donetz basin, the bulk of which were dependent on Belgian financing.

Developments within individual enterprises were outcomes of a variety of project economies and spread effects. As various components of the South Russia project took on the capabilities of a capital goods sector, two particularly dynamical processes built up, the first being the cumulative and experimental transfer of techniques through competitive emulation, the second being process and product spread effects. We might suggest, at least tentatively, that such "project effects" were crucial to the reduction of the tremendous locational disadvantages faced by any one, individual enterprise or agency of transfer.

It was undoubtedly competitive emulation that prompted Hughes to switch from the working of very impure haematites to the richer magnetite ores. New deposits were discovered monthly during the 1890s because of stiffly competitive conditions, and open working was carried to 15 or 20 fathoms as a rule. Competition drove iron extraction rates up towards 67 per cent during the late 1890s for the haematite, low silicon and low phosphorous ores, a yield comparable with the 70 per cent expected of twentieth-century operations.[44] Competition was heightened through the accumulation and movement of pools of skilled labour brought in from elsewhere. The later development of metalworking and other industries at the coastal site of Taganrog was associated with a swelling of the skilled population, especially of Belgian metalworkers. Technological experiments took place within the major enterprises, from the trials of working highly phosphorus ores with the basic process at the Briansk blast furnaces to the scaling up of blast furnace capacities at the New Russia works. The technological impacts of competition were well illustrated in the coking of coals. To 1893, Schaumburg ovens dominated the coking process. In that year model

---

[42] "Iron and steel manufacture in Russia", *The Engineer*, 26 June 1891, p. 512. In addition, works in Russian Poland manufactured zinc, cast iron, bar and rolled iron, much of which came formerly from Germany.

[43] *The Engineer*, 20 September 1901, p. 303, quoting official returns, which also enumerated 264 blast furnaces (212 hot blast) using coal, coke, and charcoal.

[44] *Revue industrielle des mines*, 17 September 1896, p. 341; *The Engineer*, 12 July 1901, p. 39.

Coppée ovens were erected, installed, and worked by the makers. By the end of 1895, the coal mining companies had installed 432 of the superior ovens, and the five largest ironworks were in addition using 842 Coppée ovens. The new technique produced at a rate of 760,000 tons of coke per year, ample with which to manufacture the 414,000 tons of pig iron then produced.[45]

By the late 1890s, project-spread effects took a variety of forms. The core mining and metallurgical enterprises were surrounded by a technological hinterland of enterprises producing sugar or foodstuffs or textiles. The early centre of Ekaterinoslav was, by then, undertaking complete locomotive manu-facture and cement and salt production, mostly under French or Belgian auspices. Newcomer locations within the project, such as Taganrog, had from the beginning a wide production profile, in this case one that included shipbuild-ing and very varied engineering workshops. The capital goods sector which was developing out of the core set of enterprises spawned throughout South Russia a multitude of enterprises engaged in flour making, non-ferrous metallurgy, and glass and cement manufacture. Indeed, as the core industries became more fully indige-nized, the more diversified areas of the project were the ones to carry further levels of foreign technological influence. Thus the fast emerging Baku oil industry was, until the mid-1890s, dependent on imports of vital equip-ment from the USA, Britain, and Germany. By 1898, engineering firms in Ekaterinoslav, Marinpol, Taganrog, and Lugansk were replacing foreign suppliers in such relatively sophisticated areas as oil-line tubing production.[46]

The success of the South Russia project during the 1890s may be interpreted in terms of project dynamics (e.g. spread effects) or transfer mechanisms (e.g. joint ventures or skill transfers). But three fairly direct and simple factors are worthy of emphasis at this point. First, the technology of the project owed much to the learning experience associated with foreign enterprise prior to 1887. Many attempts to establish modern iron working around the Donetz, using French and German artisans, preceded the key breakthroughs of John Hughes. Secondly, it is clear that the Russian government was long committed to the establishment of a Russian mining and metallurgical sector. However unsuccessful, government intervention entailed the establishment of significant administrative and bureaucratic capacity before 1887, including the development of a reasonably sophisticated Mining Department within the Ministry of Finance, a Corps of

---

[45] George Kamensky, "The ironworks of the south of Russia", *The Engineer*, 6 September 1895, pp. 241–2.

[46] The older port cities were transformed by the Project. New firms dependent mostly on foreign financing abounded in Odessa — the Belgian Société métallurgique d'Odessa, managed by an expert from the Odessa Tramways, E. Cambier, produced iron and steel from scrap. Similarly, foreign technological influences came to Nicolaiev through newly established Belgian and Russian firms devoted to ship repair and construction, steam engine and boiler manufacture, artillery and torpedo production, and so on.

Mining Engineers, and a government school of mines at St Petersburg.[47] Thirdly, and perhaps explaining much of the above, the South Russian project may not be properly understood outside the context of the long-term military and strategic requirements of the state, especially as these centred upon railways, shipping, and modern communications.

The failure of the South Russian project about the turn of the century had little directly to do with problems of technology transfer. The downturn in the entire Russian economy lasted through the period approximately 1901–6, but it is possible to identify factors more proximate to the metallurgy and metalworking sector. Western observers identified a crisis of overproduction brought on by the artificial climate of tariffs and government contracts. In this view, Witte's ministry was advised to abandon special treatment and to let the iron industry "work out its own salvation by the unfailing and well-tried process of natural selection, which should result in the survival of the fittest".[48] Similarly, a congress of Russian iron founders and metallurgical firms which met in 1900 blamed the downturn in the sector on the failure of government to continue its contracting for public enterprises, especially railways and shipbuilding. Witte's reported reply was that imports of iron and machinery continued and were a measure of the failure of the Russian industry to meet the demands of private enterprise. In this view, the problems of the industry were not those of a lack of skilled labour nor of government demand, but of a narrowness in the range of high-quality products manufactured within Russia. Witte's tentative plan further to update the technical profile of the industry by bringing in even more foreign capital was finally dashed by changes in the international economy that were beyond his or his government's control.[49]

## LESSONS OF THE FIRST CLIMACTERIC

The obvious contrasts between Japan and Russia in size, location, and frontiers, in their respective cultural, political, and commercial relations with the more

---

[47] For example, it was the experiments and geognostic surveys of the Mining Department which first showed that Leplez's famous judgement of thirty years earlier — that the coal basin of South Russia could not serve as the basis for a Russian iron industry — did not hold; both coal and iron were discovered in the Western Donetz. This made the government's experimental iron manufacturing at Lugansk viable and led to the establishment, by the government, of a blast furnace suitable for casting and wrought iron production, using coke for smelting. The whole experiment was financed by government, all the necessary machinery manufactured at government works in Lugansk. This represented the first real success at the employment of mineral fuels.

[48] "Russian iron industry", *The Engineer*, 8 February 1901, p. 150.

[49] "The Russian iron trade crisis", ibid, 27 December 1901, pp. 655–6.

advanced industrial powers, and in the political outcomes of their early industrializations, have not prevented historians from drawing lessons from their conjoint experiences.[50] Taking a fairly narrow span of years and a reasonably discrete phenomenon, industrialization through technology transfer, we may here emphasize those dynamical features that Japan and Russia had in common. Outstandingly, both represented relatively backward industrializing systems. As late as the outbreaks of the first world war or the Russian Revolution, levels of per capita income in Russia and Japan remained much lower than those of such leaders as Britain, North America, and Germany, and still significantly below those of such later followers as Austria and Italy.[51] Both nations industrialized under the auspices of the military state and both utilized a great variety of transfer mechanisms made available through the great expansion of the international economy and its institutions and the tremendous innovations in transport and communications. Both nations escaped colonization by the Atlantic powers and went on to wield their increased industrial capabilities in colonial adventures of their own.

The two systems collided in 1904–5. At the core of the Russia-Japanese war of those years was Russia's dependency upon a single modern technology, the Eastern Siberian Railway, which dictated the limits to the mobilization of armies and equipment.[52] For both nations, the war had profound outcomes in the longer term. Japan's victory meant a surge of foreign capital into the nation, an increase in the pace (and an extension of the mechanisms) of technology transfer, and a confirmation of the statist industrial programme. In Russia, defeat was an element in the civil breakdowns of 1905, the extension of recession, and the ideological turmoil that was to lead to the Revolution of 1917. From all this, it might be supposed that the cases of Japan and Russia would serve as interesting and apt examples of the late-nineteenth-century climacteric and permit some

---

[50] For the best wide-ranging attempts see Angus Maddison, *Economic Growth in Japan and the USSR* (London: George Allen and Unwin, 1969); C.E. Black *et al.*, *The Modernisation of Japan and Russia. A Comparative Study* (New York: Free Press, 1975).

[51] P. Bairoch, *The Economic Development of the Third World since 1900* trans. C. Postan (London: Methuen, 1975); Bairoch, "International industrialisation levels from 1750 to 1980", *Journal of European Economic History* (1982), **11**: see tables on pp. 281, 294, and 296.

[52] The low standard of the Russian railway, especially in winter around Lake Baikal drastically reduced the ability to mobilize: 80,000 troops were fielded against the 180,000-strong army of Japan. The battle fleet was more balanced. The debate over whether Russia could hold Port Arthur, the nation's only ice-free port on the Pacific, on the basis of a naval engagement, centred on technological capabilities. The importance of getting troops supplied and into position is easily demonstrated: on the fall of Port Arthur, Russian casualties numbered 28,000, while those of the Japanese numbered 58,000. When the Russian Baltic fleet finally arrived in May 1905, it compared badly with Togo's modernized navy, this resulting in Japan's complete naval victory at Tsushima on 28 May 1905. The estimated cost of war to Japan was around £100 million, to Russia somewhat less.

generalization as to the historical conditions of successful technology transfer and late development.

At one point during the early 1880s, the major Western newspaper in Japan complained, not for the first time, that "Japan cannot take the material, and leave the moral, civilisation of the West".[53] The principal function of the state in both Japan and Russia was to confound this view by establishing a context for limited change. The state was the only agent capable and willing to reduce the locational disadvantages of relatively backward sites by instituting systems of social control, information dispersal, demand creation, and risk reduction. All such functions lay well beyond the direct investment of the state, so often measured and centred upon by historians. Successful state activity required time and the incurring of costly failures in a period when no other agency was available. The failure of the Russian state in attempting to establish metallurgy in the 1860s and 1870s served as a positive resource which was drawn upon in the later more successful phases of transfer. In this case, the resources included accumulated administrative and technical skills, machinery and equipment, and entrepreneurial experience. In Japan, the more broadly based intervention of the state allowed it to operate at an enormous variety of levels and to establish procedures for information dis-persal and risk reduction.

Both the cases of Russia and Japan demonstrate the significance of an open system in which various transfer mechanisms co-exist and compete. With reference to points made in the first section of this essay, the manner in which the state continued to monitor, manipulate, or even create transfer mechanisms may well have depended upon crucial choices made at an earlier stage. In this sense, a trajectory of State-initiated institutional innovations may be established which continues into later phases of industrial modernization. Thus Witte's inheritance of a commitment to the importing of capital and to associated foreign enterprise led naturally enough to a coalescence, in Russia, of all manner of mechanisms of transfer. The Japanese government's essentially political and strategic decision to prohibit the use of foreign entrepreneurial capital in Japan itself meant that the range of transfer mechanisms was more restricted.[54] Consequently, the Japanese government was forced continually to widen the

---

[53] *Japan Weekly Mail*, 16 February 1884, p. 67.

[54] For details of the background to the Japanese decisions over capital, see Ian Inkster "Into the twentieth century: patterns in the relations between science, technology and the state during the early industrialisation process", paper for Round Table 4.3–4.4: "20thC science: beyond the metropolis", ORSTOM/UNESCO Conference, Paris, Unesco, 19–23 September 1994. Of course, the foreign experts in Japan were full of unsolicited advice to the Japanese government on the importance of foreign capital and enterprise. The argument tended to take either a stance straight from J.S. Mill (who was frequently quoted in the press) or argued on the basis of a need for specific expertise, as in C. Netto, *On Mining and Mines in Japan* (Tokyo: Memoirs of the Imperial University, 1879); see also *Nature* (1880), **5**: 316.

range of transfer mechanisms through varied institutional devices. The state had no choice but to provide the information and incentives system made available elsewhere by direct foreign enterprise activity.

It has been misleadingly suggested by numerous writers that the Japanese government reduced its activities generally around 1881–4 and withdrew from the transfer process as private interests took over.[55] In fact, measures such as the direct employment of foreign experts (see above) were merely replaced by a greater variety of measures, including the institution of a patent system and the later emergence of direct investment.[56] Thus, during the build up to the war with Russia, when Japan was increasingly in receipt of foreign capital, the government continually aided the expanding private sector. The government sent substantial parties of Japanese students to Europe for naval and other engineering experience; in this sense, the shipyards of the Thames, Tyne, and Clyde were among the training sites of the Imperial Navy.[57] The Japanese authorities employed foreign engineering agents to place large contracts for official orders of highly strategic equipment. The Yokohama-based merchant F.W. Horne, for example, was sent to America to place orders worth £50,000 with such firms as GEC and Morris Machine Co. to supply equipment for coastal cruisers and arsenals.[58] The government continued to act as a screen.

The process whereby the idigenous private sector took up the opportunities relayed by government also took time. The followership of private enterprise was vital in increasing competitive emulation and the efficiency of core transfers, but it was also of significance in spreading the applications of advanced techniques to other products and other locations, clearly shown in the Russian case. The prime sites of new transfers were to be found well beyond the initial production sites nominated by the state. In essence, if original transfers were stylized and statist, the emulative and competitive diffusion of foreign technology belonged mainly to the realm of private enterprise, a sector benefiting also

---

[55] This is in fact the most common viewpoint expressed on the role of the Meiji government. See as an example only R. Minami, *The Economic Development of Japan. A Quantitative Study* (London: Macmillan, 1986), pp. 154–5. Minami also argues that the emphasis on the state is part of a Marxist hypothesis (in Japan) and that the responsiveness of the private sector was of greater importance, this dictated by the availability of savings, enthusiasm and "the close relationship between the government and the private sector" (p. 155). Given the last remark, the role of government in generating both enthusiasm and savings, and its tremendous role in the production of public goods and infrastructures and services, the notion of a stark transition or of government "failure" totters on shaky ground. See also below.

[56] For patents, see Ian Inkster, "On modelling Japan for the third world (Part 1)", *East Asia* (1983), **1**: especially Section II, and more generally Inkster, *Science and Technology in History*, chapter 7.

[57] *Japan Gazette*, 14 July 1902, p. 8.

[58] Ibid., 14 July 1902, p. 9.

from the infrastructural and other impacts of earlier government activity. This dual characteristic of a time-lag and a move between areas of production and/or location (rather than a simple "taking over" from the state) is an explanation of how a national system may harbour, at one and the same time, a strongly controlled element dominated by the state and a series of innovative jumps in the private sector. With reference to the commentary in my opening section, the private sector both "follows in the wake" of the state and yet, to an extent, lies beyond it. Complete state control would possibly speed up initial transfers but would almost certainly dampen major cumulative effects.

The fourth immediate lesson of the case studies concerns competition between suppliers. This is an essential ingredient in successful technology transfer, influencing strongly the nature of the technique actually available and the process whereby it is adapted to new factor endowments and other circumstances. Any mechanisms that increase such competition may be interpreted, therefore, as mechanisms of transfer. In the absence of a foreign private enterprise sector, competition between suppliers in Japan was (until the 1890s, at least) increased through alliances between governmental agencies and foreign traders located within Japan, an arrangement almost certainly aided and abetted by a host of information dispersals operating within the heady confines of the treaty ports. Only through competition in supply could the range of advanced techniques be literally "appropriated" in either case; the alternative may well be the unsuccessful import of inappropriate techniques.

It is tempting to identify some of the more generic implications of the case studies. One appears to be that analysis of the characteristics of the principal technological donor at the global level does not offer any key to the overall pattern of technology transfer.[59] In particular, Japan did not import significant amounts of British capital. Russia financed its industrial modernization from France and Belgium rather than from Britain. In addition, in both Russia and Japan the machinery exports of Britain faced very stiff competition from those of other nations, especially Germany and the USA. Whatever the importance of Britain as the financial and trading centre of the world economy, the dynamics of the climacteric depended on something more than this.

Another implication of the case study approach centres on the importance of institutional innovations. This was perhaps more obvious in the Japanese case, where the long years of isolation had left Japan a *tabula rasa* upon which foreign cultural traits had hitherto represented *hakurai*, mere importations. The internalization of Western technology did not require wholesale cultural

---

[59] See chapter 11 of Nathan Rosenberg, *Inside the Black Box*, where complementary trade and investment relations between Britain and her principal partners are emphasized. Most investment from the United Kingdom did not go to fast-industrializing systems but to areas of recent settlement as part of a strategy of reducing the price of raw materials.

revolution but it did depend on key institutional adjustments in the fields of financing, social control, education and training and political ideology. Kranzberg has emphasized that "the transfer of technology involves a disruption of previous work habits and thought patterns",[60] but the extent of disruption depends on both the physical imperatives of the technology concerned and on the manner of its establishment or settlement. In Japan both of these factors served to reduce disruption. In Russia the first factor certainly increased the likelihood of social disruption; the second possibly acted in the same direction.[61]

Thirdly, success during the first climacteric seems to have been contingent upon military expenditure and the emergence of a military-industrial complex centred on the application of transferred modern techniques. Given their low levels of per capita income, both Japan and Russia spent a disproportionately large amount on military enterprise, particularly if the latter is taken to include much investment in transport and communications.[62] It has been argued that in more mature industrial economies, such as late-nineteenth-century Britain, military and colonial expenditures dampened the growth process, removing investment funds from areas of greater growth potential and so reducing the rate of technological change.[63] In nations such as Japan and Russia, the opposite might well have occurred.

## CONCLUSIONS: ON CATEGORIES AND DOMAINS

It seems clear that the historical phenomenon of technology transfer may not be categorized as somehow "belonging" to the study of either economic history, cultural history, or the history of technology. A proper appreciation of the

---

[60] Melvin Kranzberg, "What constitutes an industrial revolution", *Foreign Policy Bulletin* (1960), **39**: quotation on p. 96.

[61] A crucial problem for Russia was the relative decline of the *kustarni* (rural handicraft) industries which in Japan acted as a focus for diffused low-level technology transfers and provided continuing employment opportunities. The existence of a prosperous rural *industrial* sector served to reduce the socio-economic dualism associated with agriculture and industry in Tsarist Russia. See also: T. von Laue, "Russian peasants in the factory 1892–1904", *Journal of Economic History* (1961), **21**: 61–80; M.K. Palat, "Tsarist labour policy", *Soviet Studies* (1973), **2**: 62–9; G. von Rimlinger, "Autocracy and the factory order in early Russian industrialisation", *Journal of Economic History* (1960), **20**: 67–92.

[62] K.M. Hobson, "The military-extraction gap and the wary Titan: the fiscal-sociology of British defence policy 1870–1913", *Journal of European Economic History* (1993), **22**: 461–506.

[63] L.E. Davis and R.A. Huttenback, *Mammon and the Pursuit of Empire* (Cambridge: Cambridge University Press, 1986); P.K. O'Brien, "The costs and benefits of British imperialism, 1846–1914", *Past and Present* (1988), **120**: 163–200. The arguments have focussed on the rate of British productive savings.

imperatives and impacts of the Trans-Siberian Railway or the Kianguan Arsenal may well demand of the historian a knowledge of technical detail and a competence in engineering principles. But, in itself, such knowledge may not often yield an understanding of "success and failure" or a measure of the contribution of specific technological transfers to industrial and economic development. Although statements of precisely this sort, concerning the boundaries of categories and domains, are common enough, and frequently valid, in the arena of the historical interpretation of significant episodes of technology transfer they are of especial pertinence. Few historians who take up a study in this field, particularly if it concerns transfers of technology between very different technological systems, would come away with a sense that all is right with their own disciplinary perspectives, and this is true of both professional historians of technology and economic historians. In this sense at least, a study of technology transfer may all too readily become a study in history of potentially Toynbeean dimensions.[64]

It is something of a temptation, therefore, to adopt simplifying devices, the best-known of which stem from developments in economic theory. The deployment of such theoretical devices has, indeed, the effect of carving out a discrete domain of study based on operationally conceived or opportunist categories. For instance, we might make the simplifying assumption that growth through technology transfer comes about through imitation, innovation, and exploitative efforts, and then hazard direct measures of growth, imitation, and innovation levels over a given period. A simple equation linking the three "known" variables will leave "efforts" as the derivable unknown and allow the analyst to make very authoritative-seeming judgements about the contrasts between different nations in their "efforts to modernize", and perhaps go on to use such distinctions as an explanation of success or failure.[65] Explanations emanating from such devices are unlikely to posit international economic forces, colonialism, blocked transfer mechanisms, or problems of internal social control as being of significance to "failure"; rather, they are likely to end with statements as to "natural affinities" for "catch-up" or "emulation" and so on.

Such an extreme economistic approach would surely omit many of the salient features of our two case studies. Hopefully, it has been illustrated how the

---

[64] This is not meant to be at all facetious. It is quite possible to interpret Toynbee's massive work in terms of challenge and response, within which the transfers of technology are quite central. There is, indeed, quite a lot of technology in Toynbee.

[65] The example is not entirely hypothetical. See Jan Fagerberg, "Why growth rates differ", in Giovanni Dosi *et al* (eds.), *Technical Change and Economic Theory* (London and New York: Pinter, 1988), pp. 432–57, especially pp. 438–40, where this "theory" is proposed. Claiming to be novel because it "incorporates the effects of national innovative performance" (p. 440), the device unravels with the proxy for the "total level of knowledge appropriated" (the level of productivity) falling to real GDP per capita. The proxy for a nation's own "creation of technology" becomes (for twentieth-century nations) R and D expenditure. So many slips between the cup of theory and the lip of history.

growth of knowledge transferred into Japan and Russia during their industrial drives may be indicated and judged but not quantified in such a manner that accurate or seeming "scientific" cross-comparisons may then be made and relied upon for further refined deductions. Even the bench-mark income per head levels of such systems are very seriously in doubt among more thoughtful economic historians, who may not then simplify their arguments by assaying a growth rate for an economy. When the growth figure itself is fuzzy and when yet other variables are proxied (for instance, the increase in technical capability within a receiver system might be proxied by figures on R and D expenditure), it becomes certain that such approaches contain little historical material, no sensitivity, and therefore no empirical check on the theorizing process.

It should also be obvious, however, that a hidebound historicism or disciplinary nervousness is as unlikely as abstract economics to result in sturdy and attractive interpretations. Goethe may or may not have believed that "everything factual is already a theory", but historians who are attracted to that view must still make some informed decisions about when to stop. Neither abstraction nor a perverse facticity are able to produce reliable historical stories. In particular, episodes of massive technology transfer represent time with an irregular beat, even explosive time. There is not yet any formula manufactured which catches the dynamics of the conjuncture.

# The Japan That Can Say No: The Rise of Techno-Nationalism and its Impact on Technological Change

*Morris F. Low*

Historical accounts of Japanese technology are difficult to find. Scattered throughout many disciplines, they lack a clearly defined methodology and core literature.[1] Given the different concerns of the writers, using this diverse literature can be fraught with difficulties, but historians should by no means dismiss what are often accounts of corporate success written for the masses. Some of this literature has pointed to a "sea change" in the way the Japanese view their national interest and techno-security, apparently arising from the nuclear threat posed by North Korea, the instability of the former Soviet Union, and growing anti-US sentiment.[2] How will this affect technological change and the way we account for it?

---

[1] Notable monographs recently published in English on post-war Japanese technology: Commission on the History of Science and Technology Policy, *Historical Review of Japanese Science and Technology Policy* (Tokyo: Society of Non-Traditional Technology, 1991); M.A. Cusumano, *The Japanese Automobile Industry. Technology and Management at Nissan and Toyota* (Cambridge, Mass.: Council on East Asian Studies, Harvard University, 1985); S. Nakayama, *Science, Technology and Society in Postwar Japan* (London: Kegan Paul International, 1991); D.I. Okimoto, *Between M.I.T.I. and the Market. Japanese Industrial Policy for High Technology* (Stanford: Stanford University Press, 1989); R.J. Samuels, *The Business of the Japanese State. Energy Markets in Comparative and Historical Perspective* (Ithaca: Cornell University Press, 1987). For the period before the second world war, see C. Kamatani, "The history of research organization in Japan", *Japanese Studies in the History of Science* (1963), **2**: 1–77.

[2] R.S. Morse, "Techno-nationalism on the rise", *Tracking Japan. A Washington Report on Trade, Technology and Japan* (April–May 1994), **1**, no. 4: p. 3.

To understand how it might, this paper will examine the book *The Japan that Can Say No*, written by the politician Shintarô Ishihara and the chairman of Sony Corporation, Akio Morita. Although not a rigorous account of the role of technology in bilateral relations, the book nevertheless does throw light on the phenomena of techno-nationalism and the authors' views on how Japan succeeded and what it needs to do in the future.

The diversity of books on Japanese technology reflects the broad interest in the relations between science, technology, and Japanese economic growth, but it is also evidence of the general malaise that has befallen the history of science and technology, and the fledgling state of STS studies in Japan.[3] Given this situation, perhaps we should be grateful that a literature is being generated by the business community, albeit often with what the Japanese historian of science Shigeru Nakayama describes as the "business school approach". Accounts are written like *The Japan that Can Say No* which unabashedly reflect national interests and business biases rather than any "objective" position to which an historian might aspire in his or her work.[4]

But are such national boundaries still valid? In writing a post-modern history of Japanese technology, we need to reconsider the extent to which geographical boundaries determine the dynamics of the spread of Western technology and to ask whether the idea of a region is sometimes more useful and how far a nation's borders define a national technological system. Perhaps the way in which many historians, and certainly many businessmen, have mapped the development of technology in terms of nation-states needs to be reexamined. Certainly the post-Cold War transformations occurring in Europe tell us that this will be necessary in the future.

My discussion of these questions is divided into three sections. The first examines the threat which, as I argue, is presented by techno-nationalism, as discussed in the book *The Japan that Can Say No*. In the second, I ask, if there is some basis for concern, to what extent it will affect technological change in the future, bearing in mind that the reinforcement of national borders has been countered by the recent globalization of R&D which has seen a leakage from national technological systems to an international space. In the third and final section, I discuss the implications of transnational activity for our understanding of Japanese technology and the writing of its history.

---

[3] M. Fridlund, *The Teaching of History of Technology in Japan. A Survey in 1990* [Trita H.S.T. Working Paper no. 93.7] (Stockholm: Department of History of Science and Technology, Royal Institute of Technology, 1993).

[4] S. Nakayama, "History of East Asian science: needs and opportunities", *Osiris* (1995), **10**: 80–94.

# THE JAPAN THAT CAN SAY NO

At the beginning of the new Heisei era in January 1989, a small book of 160 pages was published by Kôbunsha, one of Japan's larger publishers. Its title was *"No" to ieru Nihon: Shin Nichi-Bei kankei no kaado (The Japan that Can Say No. The new card to be played in US–Japan relations)*. An unofficial translation of the book, produced by the Pentagon's Defence Advanced Research Projects Agency, caused such fierce controversy that part of the "bootleg" English version was read into the minutes of the US Senate. An official translation has recently appeared, but the coauthor Akio Morita has refused to allow his portions of the original book to be included, in order to avoid any further misunder-standings and to limit the commercial impact on Sony arising from flak from the book.[5]

*The Japan that Can Say No* raises the issue of the rise of "techno-nationalism", both in Japan and the USA.[6] What implications does this have for the history of technology and for studies of Japanese industry? In the new world order, the Soviet threat may have abated, but the "technonationalism" reflected in *The Japan that Can Say No* has not gone away. The tendency on the part of nations such as the USA to withhold commercially sensitive scientific and technical information from Japan and other nations has been partly countered by the so-called "internationalization" of Japanese science and technology. This has taken the form of an attempt to maintain an adequate inward flow of know-how by the sending of students and employees overseas, by the establishment of think-tanks abroad, and US–Japan science and technology agreements which facilitate the transfer of military technologies.

It is unfortunate that such a hastily produced book has attracted so much attention. The book was published as a volume in Kôbunsha's "Kappa Books" series which is aimed at the mass market, and it sold at the low price of 820 yen. The character of the book, too, was populist, its content being transcripts of speeches given by Shintarô Ishihara, a long-time member of the Liberal Democratic Party (LDP) and a member of the Japanese Diet, and Akio Morita. The audience consisted of Ishihara's supporters in his constituency of Shinagawa, where, coincidentally, Sony headquarters are located.

---

[5] S. Ishihara, *The Japan that Can Say No. Why Japan will be First among Equals* (New York: Simon and Schuster, 1991).

[6] For a discussion of the rise of techno-nationalism in Japan, see S. Nakayama, *Shimin no tame no kagakuron [Studies of Science for the People]* (Tokyo: Shakai Hyôronsha, 1984), pp. 187–211.

Like Morita's previous book, *Made in Japan. Akio Morita and Sony* (1987), which relied on journalists Edwin M. Reingold and Mitsuko Shimomura for editorial assistance, *The Japan that Can Say No* has relied on the skill of Kôbunsha's staff to turn a string of anecdotes and rhetoric into a publishable manuscript. As Morita himself has recently said, "Many books like that are done for politicians".[7] The book has proved to be a great fundraiser for Ishihara, the controversy in the United States helping it to sell over one million copies in Japan.[8]

Despite the problem of its being written for a Japanese audience, the book is significant in that it does reflect opinions held by many Japanese regarding US–Japan relations and American racist attitudes; cultural explanations for Japan's economic success; the trade imbalance and the shortcomings of US business; the rise of Japanese technology as a strategic bargaining card, and Japan's need to say "no" to the United States when economic and political considerations make this desirable. While this last point comes out strongly in the book, such attitudes and issues have been articulated by Morita, Ishihara, and other Japanese writers elsewhere.

Akio Morita was one of the founders of Sony and, as such, is thought by the general public to be vested with special "insights" into Japan's post-war success. Unfortunately the insights tend towards well-worn stereotypes, assuming the character of reminiscences by a successful businessman rather than a systematic analysis of Japan's economy. In both books, Morita tries to explain Japan's success as the result of unique qualities, such as the Japanese devotion to work and the Japanese loyalty to their company, which take priority over monetary concerns. Morita comments on how the humane Japanese never fire workers because of the family bonds created within companies. In reality, this is no longer true in recession-hit Japan. Morita also forgets to mention that only about one third of the Japanese workforce is covered by the lifetime employment system.[9]

Anyone who has read Morita's *Made in Japan* would hardly be surprised by the arguments advocated in the chapters bearing his name in the original Japanese version of *The Japan that Can Say No*. In *Made in Japan*, he argues that:

> ... if you are trying to motivate, money is not the most effective tool ... you must bring them into the family and treat them like respected members of it. Granted, in our one-race nation this might be easier to do than elsewhere, but it is still possible if you have an educated population.[10]

---

[7] "Morita: 'A turning point'", *Business Tokyo*, February 1990, p. 30.

[8] See Kappa Books advertisement, *Asahi Shinbun*, 4 June 1990, p. 3. Sales have exceeded 1.1 million copies.

[9] I. Buruma, "We Japanese", book review in *The New York Review of Books* (12 March 1987), **34**, no. 4: 16–18. For a further rebuttal of Morita's arguments, see M. Godet, "Ten unfashionable and controversial findings on Japan", *Futures* (August 1987), **19**, no. 4: 371–84.

[10] A. Morita with E.M. Reingold and M. Shimomura, *Made in Japan. Akio Morita and Sony* (London: Collins, 1987), p. 138.

In the Japanese case, the business does not start out with the entrepreneur organizing his company using the worker as a tool.[11]

In both books, Morita accuses Americans of fickleness in their dealings with the Japanese:

Why are congressmen and others complaining about Japanese competition when American automobile companies themselves are increasing their purchases and even complaining that they cannot get enough cars?[12]

Then, as a prelude to the book to come, he argues that:

If Westerners and Japanese are ever going to understand each other, the Japanese will have to be as frank as the Americans are in discussing problems and putting forth their viewpoints. We have been very poor at this in the past, businessmen as well as politicians, and we do not seem to be learning fast enough.[13]

A businessman (Morita) and a politician (Ishihara) did get together, and their outspokenness stirred up a controversy. Morita's line of argument — "that America is very much to blame for her problems, not just Japan"[14] — was one of the aspects that upset American readers. Despite his plea of naivety — that the Japanese language version of *The Japan that Can Say No* was not intended to be picked up by an American audience — he did confess that "I try to say things that will get people's attention and get them thinking clearly about our problems".[15] A little later in the book, he wrote: "We never seem to be holding our heads up high and saying sincerely just what we think".[16] Ishihara provided Morita with the perfect partner.

We can perhaps forgive Morita for voicing such stereotyped claims since many Japanese entertain the same simplistic interpretations of their own society; also such ideas have appeared elsewhere in his writing, and serious Japan scholars know better than to take what he writes literally. Rather, it was Ishihara's assertions that stirred up the controversy. He suggests that American attitudes to the Japanese are racist, and that Japan could sell its technology to what, at the time he wrote, was still the Soviet Union. Technology provides the Japanese with an ace card with which to threaten America if it is bullied too much. Such an irreverent attitude to the US–Japan alliance was not new. Ishihara is a successful novelist who entered politics in 1968. Shortly afterwards, in March 1969, he made known his view that Japan should not rely on the United

---

[11] Morita, *Made in Japan*, p. 143.

[12] Morita, *Made in Japan*, p. 225.

[13] Morita, *Made in Japan*, p. 264.

[14] Morita, *Made in Japan*, p. 290.

[15] Morita, *Made in Japan*, p. 290.

[16] Morita, *Made in Japan*, p. 291.

States for defence and that Japan should arm itself with nuclear weapons.[17] Ishihara has subsequently held the position of Minister for Transport, and in August 1989 was an unsuccessful candidate for the presidency of the L.D.P. and the position of prime minister. His literary aspirations continue and he heads a group that holds a special remembrance every year for Yukio Mishima, who committed suicide in November 1970 after a failed attempt at persuading the Self-Defence Force to take up arms in the name of the Emperor.[18]

Given his right-wing views, it is not surprising that the development of the next-generation support fighter, code-named FSX (fighter support experimental) and designed by Mitsubishi Heavy Industries, was so close to his heart. He was alarmed that Prime Minister Yasuhiro Nakasone gave in to US pressure and had agreed to develop and produce the fighter as a joint project. These views come out in a discussion with the American journalist James Fallows in the January 1990 issue of the general-interest monthly magazine *Bungei Shunju* (circulation in 1990 of 750,000).[19] Ishihara voiced his opinion regarding the FSX project elsewhere as well. In an article that originally appeared in July 1989 in another prestigious monthly magazine, the conservative *Chûô Kôron*, Ishihara complained that Japan was sold short. Despite Japan's being able to develop the FSX on its own, a decision was made to treat it as a joint development project with the United States. The special characteristic of the fighter was that it would be able to circle quickly with about one third the turning radius of F-15 and F-16 fighters. According to Ishihara, under the agreement, Japan virtually agreed to supply all of its own relevant aviation technology to the USA free of charge, whereas it had to pay for the American technology. The USA would be able to use the Japanese technology freely; the Japanese, however, were unable to use the American technology for purposes outside the FSX project. Despite this, the Japanese have been accused of trying to steal American technology via the project, in what Ishihara calls a display, on the part of the USA, of "techno-nationalism".[20]

The renegotiated FSX agreement included a number of controversial aspects. American firms would in the future be allocated 40 per cent of the project work; the Japanese would be denied access to US flight control software; the USA would, however, have access to technologies, such as carbon-fibre technology,

---

[17] J. Welfield, *An Empire in Eclipse. Japan in the Postwar American Alliance System. A Study in the Interaction of Domestic Politics and Foreign Policy* (London: Athlone Press, 1988), pp. 260–1.

[18] "Taiketsu! Nichibei no byôkon o tataku" ["Showdown! Thrashing out the cause of US–Japan problems"], *Bungei Shunju,* January 1990, 148–60. Fallows was the author of a controversial article entitled "Containing Japan" which was published in the May 1989 issue of *Atlanta Monthly.* For a critique, see K. Nukazawa, "Flaws in the 'Containing Japan' thesis", *Japan Echo* (1989), **16**, no. 4: 52–7. Fallows has discussed the idea of imposing a 25 per cent flat tax on Japanese imports in the December 1989 issue. See Chalmers Johnson, "Revisionism and beyond", *Business Tokyo,* February 1990, 54–5.

[20] An abridged translation appears as S. Ishihara, "From bad to worse in the FSX project", *Japan Echo* (1989), **16**, no. 3: 59–65.

developed in Japan for the project; and US military patents will be safeguarded by Japan's Defence Agency, not its Ministry of International Trade and Industry (MITI).[21]

In an article in the March 1989 issue of *Bungei Shunju*, Ishihara expressed the view that:

> Creating its own weapon systems using its own leading-edge technologies would ... display in a very dramatic fashion Japan's latent power in international politics and give Japan greater capacity to resolve world tensions and compel the military superpowers to compromise and cooperate with Japan.[22]

In Ishihara's view, Japan should not necessarily ditch the US–Japan Security Treaty, but observation of the treaty should, it seems, be secondary to the need for Japan to be responsible for its own defence. Japan, in fact, should use advanced technology as a weapon for nationalistic purposes, to enable it to be heard in international politics.

Ishihara was not alone in advancing these arguments. In early May 1989, after the conclusion of the talks over the joint development of the FSX fighter, editorials in influential newspapers, such as the largest business and financial newspaper *Nihon Keizai Shinbun* (circulation in 1991 of 3 million) and two of the three largest national newspapers, the conservative *Mainichi Shinbun* (4.1 million) and more liberal *Asahi Shinbun* (8.3 million), joined him in expressing concern over the results of negotiations. The agreement was seen in these editorials as bringing Japan little benefit and causing anti-American sentiment among Japanese.[23]

The strong feelings aroused by the FSX agreement and *The Japan that can say No* created enough interest to warrant publication of an official translation of the book by the New York publishing house Simon and Schuster, minus the chapters by Morita. Given the public response, Ishihara teamed up with Shôichi Watanabe and Kazuhisa Ogawa to produce a sequel entitled *Soredemo "No" to ieru Nihon. Nichi-Bei kan no konpon mondai (Even so, Japan can say No. Basic Problems of the US–Japan Relationship)*.[24]

Ishihara is proud of the Japanese defence industry, and his books reflect it. This attitude can also be seen in government policy. The Defence Agency, for example, is keen to buy Japanese products, and, as a result, all major manufacturing companies in Japan have become involved in supplying military

---

[21] R. Katayama with M. Thomas, "Japanese civil aviation in the soaring nineties: will it fly?", *Business Tokyo*, February 1990, 18–24, on p. 20.

[22] Cited in S. Otsuki, "The FSX problem resolved?", *Japan Quarterly*, January–March 1990, 70–83, on p. 72.

[23] Cited in Otsuki, "The FSX problem resolved?", pp. 81–2.

[24] S. Ishihara, S. Watanabe, and K. Ogawa, *Soredemo "No" to ieru Nihon: Nichi-Bei kan no konpon mondai [Even so, Japan can say No. Basic Problems of the US–Japan Relationship* (Tokyo: Kôbunsha, 1990). First published on 30 May 1990.

products. Even if the initial know-how is from overseas, ships and aircraft have been made in Japan under licence by companies such as Mitsubishi Heavy Industries and Kawasaki Heavy Industries. Weapons systems, on the other hand, have been purchased mainly from the USA, but in the future, Japanese advances in semiconductor technology and other hi-tech areas will mean that in the areas of targeting, guidance, communications, and manufacturing techniques, Japan will be taking over the lead.[25]

In 1983, a transfer agreement for military technology was made between the USA and Japan; this related especially to the exchange of electronics technology with military applications from Japan to the USA.[26] In April 1988, a US–Japan agreement was reached which would allow the transfer of information on classified US military technology to Japan.[27] Later in 1988, a new science agreement was signed which included provisions to safeguard military research and prevent its transfer to other countries.[28] The concern on the part of the USA to maintain access to sensitive technologies with military applications was confirmed by an NHK (Japan Broadcasting Corporation) survey which showed that one in three electronic companies in the USA which it surveyed (a total of 30) use Japanese electronics in military products.[29]

Given these attempts by the United States to secure strategic Japanese technology,[30] it is understandable that the USA was outraged by the possibility that the Japanese would contemplate diverting defence technology to the Soviet Union during the Cold War. This is especially so when one considers the growth of the Japanese defence industry at the time. On 29 March 1990, the International Peace Research Institute in Stockholm announced the top hundred arms manufacturers in the world. The list included six Japanese companies, which, while being a seemingly small number, is large given that Japan bans the export of military technology. Should it decide to lift the ban and pursue overseas markets, it is expected that the defence industries in Britain, France, and the USA will decline.

Increasing competition in defence industries will inevitably lead to greater secrecy regarding technology. The US Department of Defence has adopted the procedure of placing the critical flight control software called "source codes",

---

[25] M. McIntosh, *Japan Re-armed* (London: Frances Pinter, 1986), pp. 52–3.

[26] McIntosh, *Japan Re-armed*, p. 51.

[27] D. Swinbanks, "Classified US technology for transfer to Japan", *Nature* (21 April 1988), **332**: 669.

[28] Ibid. See also A. Anderson, "New US–Japan science agreement reached", *Nature* (7 April 1988), **332**: 475.

[29] P. Hartcher, "The arsenal ready and waiting in the Pacific", *The Sydney Morning Herald*, 6 April 1988, 17

[30] See the section entitled "Tapping the technological potential of Japan", in R. Drifte, *Japan's Foreign Policy* (London: Routledge and Royal Institute of International Affairs, 1990), pp. 41–3.

for both the F-15 and the FSX fighter in "black boxes".[31] This prevents the Japanese from gaining access to it and from applying the know-how to their own commercial aircraft. It is part of a trend which even Americans have come to describe as "techno-nationalism": a desire on the part of the USA to make it more difficult for foreign nations to take advantage of American technology.[32]

According to a recent survey conducted by the Science and Technology Agency of Japan, the Americans are succeeding in restricting access. A total of 1,500 researchers working in the four areas of (1) life sciences, (2) materials sciences, (3) information and electronic sciences, and (4) ocean and earth sciences were surveyed. Researchers in national research institutes and universities, as well as in private enterprise were covered by the survey.[33] There was a response rate of 60 per cent; 40 per cent of the responses came from universities, with a preponderance of replies emanating from those working in the basic sciences. One in four respondents felt that the exchange of scientific information with foreign researchers (especially in the USA) had become more difficult. This was even more apparent in the private sector, where one in three researchers answered this way. In the information and electronic sciences, the figure rose to 40 per cent. Thirty per cent of respondents (and half of those working in the private sector, in the information and electronic sciences) answered that the exchange of information had become more difficult.[34]

*The Japan that Can Say No* is further recognition of the fact that the USA is no longer a hegemonic power and that the Japanese will be able to exert more influence in the future. Whether this will be through a path of technomilitarism is anyone's guess, but what is clear is that the transfer of critical science and technology is likely to become more difficult. The next section looks at the extent of the change that we can expect.

## GLOBALIZATION OF R&D

In today's global economy, technology transfer no longer follows a simple linear model. This is particularly so in the case of Japan, which has previously been accused by politicians and industrialists of avoiding meaningful technology transfers, particularly to the newly industrialized economies of Asia. It has been argued that Japanese firms tend to rely on Japanese technical experts for a longer period than American or European companies operating offshore production facilities. Consequently, the Japanese have been accused of being unwilling to

---

[31] Katayama, "Japanese civil aviation in the soaring nineties: will it fly?", p. 20.

[32] See B. Johnstone, "Rise of techno-nationalism", *Far Eastern Economic Review,* 31 March 1988, 58–65.

[33] Kenkyûsha kôryû kibishisa fuyasu Bei" ("America toughens up on researcher exchange"), *Asahi Shinbun,* evening edition, 14 May 1990, 5.

[34] Ibid., p. 5.

transfer technology and skills. Japanese companies, it is claimed, only license the technology of older products and processes, such as that of the dot-matrix printer, whereas that of laser printers is not made available. Even in the case of joint ventures, there is often insufficient information disclosed to the other partner to make any meaningful transfer of technology. In the case of the manufacture of televisions and cameras off-shore, technology transfer is avoided by procuring parts from local subsidiaries of Japanese firms.[35] Such networks ensure that the competitive edge of Japanese products — which is based on the advantages of new technologies that accrue to the owner of the know-how — is maintained. Wide access to these technologies greatly diminishes the potential profits. This is different from "mature" technologies which are more readily transferred.

If we look at the activities of multinational firms in Asia, American and European firms appear to be more willing than their Japanese counterparts to give local employees strategic positions within their operations. It could be argued that this reflects not so much a desire on the part of the Japanese to avoid the transfer of know-how as a manifestation of how the Japanese seniority system works against giving such positions to local staff. The "Japanese type" of technology transfer apparently places a great deal of emphasis on training local engineers and operators and, it is claimed, requires a large number of Japanese technical experts and more time. The continuing presence of these experts is said to allow further on-the-job training and to ensure that local employees have an understanding of the processes beyond basic operations, especially important in the event of sudden changes to model design and production methods.[36]

With the strong yen and the shift of hi-tech manufacturing offshore from Japan, the transfer of actual know-how for the production of computer parts, telecommunications equipment, and VCRs may become a reality. Electronic parts are no longer supplied from Japan alone, but come from throughout the world. Sony, Matsushita, and Toshiba, for example, secure parts from Southeast Asia. Sharp manufactures 85 per cent of its overall production overseas.[37] Hitachi, JVC, and other firms now reimport VCRs made abroad. Japan is similarly likely to become a net importer of colour television sets in 1993, the sets assembled overseas coming mainly from Malaysia, South Korea, and China in descending order.[38] What this suggests is that national borders are becoming less relevant and that techno-nationalism will be more difficult to enforce. Arguments for the importance of national culture in Japan's success become more tenuous as well.

---

[35] D. Normile, "Japan holds on tight to cutting-edge technology", *Science* (15 October 1993), **262**: 352.

[36] K.W. Thee, "Technology transfer from Japan to Indonesia", paper prepared for the Second Conference on the Transfer of Science and Technology, 3–6 November 1992, Kyoto, Japan.

[37] "Japanese high-tech manufacturing is on the move — out of Japan", *Tokyo Business Today*, October 1993, 44–6.

[38] "Imports of color TVs jump 56% in July", *The Japan Times*, 4 September 1993, 7.

The internationalization of Japan's R&D effort — the search for new ideas and ways of commercializing them — has seen the establishment of major R&D centres in Western countries: e.g. Canon in France, Hitachi in California, Kobe Steel in Surrey, Kyôcera in Washington, Matsushita Electric in Frankfurt and San Jose, Mitsubishi Electric in Boston, NEC in Princeton, Nissan in Detroit and Bedford, and Sharp in Oxford.[39] Such strategies are not unique. More than 10 per cent of the research of US private companies is conducted overseas, whereas the figure for Japan is about 2 per cent. Japanese R&D staff working in Japan for US firms represent approximately 1 per cent of the total US private sector spending on R&D.[40] Such figures do not, however, necessarily reflect the proportion of information flows between countries. Japanese R&D facilities in the USA engage in more joint arrangements with American research organizations, whereas R&D facilities of American companies in Japan tend to limit their linkages to laboratories in universities that are staffed by university professors, as a way of recruiting the best graduates.[41] But it would be simplistic to portray such operations as one-way streets. Owing to the recession and lack of capital in Japan, for example, joint ventures between Japanese and European firms have seen a flow of marketing, design, and production expertise from Japan to Europe, rather than cash.[42]

Furthermore, American access to Japanese R&D and *vice versa* is only one small part of the story. Around 80 per cent of the Japanese researchers and engineers who have gone abroad have gone to developed countries, especially the USA. On the other hand, 90 per cent of foreign researchers and engineers in Japan come from developing countries.[43] This globalization of company research confounds simplistic attempts to restrict information flows. But we do need to see it in perspective. Less than 20 per cent of the total technological activities of most highly industrialized nations occur outside the home country, indicating that most national technological systems are still largely confined within their own borders.[44]

The breaking down of national barriers that has occurred has also enabled the transfer of reject technology and nuclear waste. This is useful for industrially

---

[39] D. Kahaner, "Major western R&D centers of Japanese companies", *Kahaner Report*, 10 April 1993, unpublished electronic version.

[40] J. Sigurdson, "Internationalising research and development in Japan", *Science and Public Policy* (June 1992), **19**, no. 3: 134–44.

[41] M.G. Serapio, Jr., "A comparative study of US–Japan cross-border direct R&D investments in the electronics industry", *Kahaner Report*, 5 March 1994, electronic version.

[42] C. Rappaport, "Japan to the rescue", *Fortune International* (18 October 1993), **128**, no. 9: 30–2.

[43] National Institute of Science and Technology Policy Second Policy-oriented Research Group, A. Nishimoto and H. Nagahama, "The interchange of researchers and engineers between Japan and other countries", *NISTEP Report*, no. 16 (March 1991).

[44] D. Foray and C. Freeman (eds.), *Technology and the Wealth of Nations. The Dynamics of Constructed Advantage* (London: Pinter, with the OECD, 1993).

advanced nations, such as Japan, where, as a result of strong public opposition, it has become increasingly difficult to begin the construction of new nuclear power plants within the country. Mitsubishi Heavy Industries, one of the leaders in the atomic energy industry, has therefore chosen to expand business overseas, especially in the developing countries of Asia.[45] While environmental groups have successfully opposed the further development of atomic energy in "industrialized" countries, nuclear plants are perceived or presented in Asia as the "wave of the future". China, Indonesia, South Korea, Taiwan, Malaysia, and Thailand are just some of the Asian countries that have expressed interest in establishing nuclear power plants within their borders.[46]

# WRITING HISTORIES

What bearing does all this have on the history of technology? In this paper, I have largely discussed cases in which technology is knowledge rather than artefact. With R&D investment starting to exceed capital investment in Japan, this balance may become even more marked in the future. The corporation has begun to move from being a place of production to a place for thinking. Potential competitors in an industrial sector may now emerge from a totally different sector, and innovation can be the result of the fusing together of hitherto unrelated technologies rather than the fruits of a single technical breakthrough.

Fumio Kodama, a leading policy analyst formerly with the National Institute of Science and Technology Policy in Tokyo, suggests that this constitutes a paradigm shift, and calls for a change of focus from technical change to institutional change. He argues that the former occurs only after social institutions have adapted to the potential of the new technology. In our fuzzy, post-modern world, the old linear model of technological innovation in which know-how flowed from the USA to Japan is no longer valid. International co-operation in science and technology renders national borders less meaningful.[47] The diffusion of a technology may only be possible "after a period of change and adaptation of social institutions to the potential of the new technology".[48] The development of much new technology incrementally, and in a wide range of institutions, makes writing its history extremely difficult.

Historians of technology can learn much from business histories and the reminiscences of entrepreneurs, but they should avoid the trap of writing success

---

[45] T. Sato, "Mitsubishi Heavy Industries makes a comeback", *Tokyo Business Today,* January 1990, 50–2.

[46] N. Oishi, "Asia demands fuels N-plant competition", *The Japan Economic Journal,* 2 June 1990, 17.

[47] F. Kodama, "Changing global perspective: Japan, the USA and the new industrial order", *Science and Public Policy* (December 1991), **18**, no. 6: 385–92.

[48] Ibid., p. 388.

stories revealing the "secrets" of economic growth of the type discussed in the first section of this paper. Books like *The Japan that Can Say No* can be of use insofar as they provide evidence of different accounts of Japan's success. Like Lee Iacocca's story of his salvation of Chrysler,[49] such narratives articulate more than just "what happened". They often present a separate agenda of self-promotion and cultural nationalism.

Just as historians can learn from business studies, so MBA students, too, can benefit greatly from viewing technological change as the result of complex processes that include technology transfer. Understanding these processes today requires us to look at the question of techno-nationalism and enforceable boundaries. Even if we go back further in time and study the introduction of Western technology in the Tokugawa period and later, we need to look not only at Europe, but also at other Asian countries such as China, India, and Korea as recipients of Western technology. This will allow us to understand better the dynamics of the transfer of foreign know-how to Japan.[50] Western know-how did not always go straight from sender to addressee. Instead, it was mediated by various local cultures and regional interests, and adapted to suit each country's needs.

Given these dynamics, the tendency to attribute features of a nation's science and technology to the influence of "traditional" culture becomes increasingly problematic. Customs and habits may owe more to a region than to the space confined within national borders. Even when trends can be identified, the speed of technological change often renders them short-lived. Daniel Okimoto has argued that while Japan's strength in technology can be traced to such factors as the applied nature of its R&D, incremental improvements, emphasis on commercial applications, quality control, and miniaturization, these characteristics may be not so much national attributes as current features of its technology, which may be transitory and destined to become less prominent in the future.[51]

The desire to gain access to foreign technology, the need for more personnel and funding than can be found domestically, shared research interests, and the desire to expand research activity have led to the growth of joint ventures and international research collaborations.[52] Writing the history of joint R&D projects conducted by teams of researchers across different nations poses very real logistical problems, not the least of them being the preservation of business archives.

---

[49] L.A. Iacocca with W. Novak, *Iacocca. An Autobiography* (New York: Bantam Books, 1984).

[50] T. Nakaoka, "The European industrial economy and endogenous development in Asia", Paper presented at the Second Conference on the Transfer of Science and Technology, Kyoto, 3–6 November 1992, p. 27.

[51] Okimoto, "The Japanese challenge in high technology", pp. 541–67.

[52] J. Warnow-Blewett and S.R. Weart, *A.I.P. Study of Multi-institutional Collaborations, Phase I: High Energy Physics, Report No. 1: Summary of Project Activities and Findings, Project Recommendations* (New York: Center for History of Physics, American Institute of Physics, 1992).

International collaborations often enjoy a limited life-span. How, then, do we write the history of technology in the age of the globalization of R&D, shifting centres, and changing peripheries?

All this not only sets an agenda for studies of recent technology but also reminds us that when we describe technology as an artefact "made in Japan" or "born in the USA", we are using labels that hide the significance of the technologies as forms of knowledge that sometimes owe less to national characteristics than to negotiation, bartering, and flows of information. Japan and the USA can attempt to say no, but there are strategies and complexities in place which make the flow of know-how difficult to police. The history of technology and technology transfer can become richer by being aware of the multiple factors impacting upon the technology making the border crossings. We need also to be aware that there are varying levels of technology, such as the mature dot-matrix printer versus the leading-edge laser printer, which can be transferred. Second-hand technologies and technological waste can be transferred to nations just as easily or more easily than that which is at the cutting edge. By doing so, we will be well on our way towards producing post-modern[53] histories of technology which transcend the ideas of national borders and advocate multiple approaches to understanding the production of knowledge, techniques, and machines in an international space — a multicultural space where the role of traditional, national culture becomes blurred and less relevant.

---

[53] For a useful discussion of postmodernism, see D. Hebdige, "A report on the Western Front: postmodernism and the 'politics' of style", in F. Frascina and J. Harris (eds.), *Art in Modern Culture. An Anthology of Critical Texts* (London: Phaidon, 1992), pp. 331–41.

# Politics and the Passion for Production: France and the USSR in the 1930s

*Yves Cohen*

The thirties were highly political years in both France and the Soviet Union. In France, the Great Depression began in 1931, and the French economy did not start to take off again until 1938, following seven years in which the drive for industrial modernization slowed down. For part of this period, from 1933, the fear of fascism dominated political life, and that fear was only partially alleviated by the victory of the Popular Front in the parliamentary elections of 1936. There-after, the idea of an "économie dirigée" — a guided economy — well articulated since 1919, became more familiar. In the USSR, the first five-year plan had been launched much earlier, in 1928, at a time when Stalin's power was no longer in question. The subsequent decade revealed the methods by which he sought to reinforce his power, culminating in the great purges that occurred in 1937–8. From the economic point of view, the main element in Stalin's programme was the implementation of a totally centralized economy. His overriding aim was to suppress the slightest sign of opposition throughout the country and in every sphere of activity: political, economic, cultural, military, and so forth. In these circumstances, in both countries, production managers, by whom I mean, for the West, the managers of firms and, for the Soviet Union, the directors of production plants, could not but show an interest, and even participate, in politics in a manner quite different from that which had been normal during the twenties. This engagement with politics had a big influence on industrial life and on the process of innovation.

It is possible to distinguish three areas within which politics raised matters of concern for those I term production managers. These concerned: first, social relations within the factory itself; secondly, the extent to which the state was involved in the life of the branch of industry concerned; and thirdly, the general conception of economic and, more especially, industrial life. The last of these

215

areas came to the fore in France during the thirties, while the political action of production managers in the USSR became increasingly restricted to the first area, that of the factory.

I begin by discussing a general hypothesis on the relations between production management and politics. I then try to evaluate what kind of power the Soviet production manager enjoyed, before sketching an outline of the production managers' politics. Finally, I offer some reflexions on the relationship between what I term the passion for production, and innovation and politics.

## DOES THE NATURE OF PRODUCTION DETERMINE THE COMPANY POLITICS?

In the postscript which he added in 1984 to his famous article of 1970 "Between Taylorism and Technocracy", Charles Maier observed that in Germany and Italy "the electrical and chemical industries often became identified as special supporters of National socialist and Italian fascist policies in the 1930s". On the other hand, "directors of the traditional iron and steel firms and the heads of the manu-facturing concerns based on steel certainly proved happy enough to benefit from government contracts and autartik protection, but they lost political influence".[1] Maier developed this kind of technological determinism in his analysis of the managers' political stance by distinguishing between the old, more labour-intensive industries and the newer, more process-oriented ones. The former, according to Maier, maintained traditional, conservative positions. Their socio-political habits, formed during the nineteenth century, were fashioned by their encounters with their organized workforces, and they relied upon patriarchal, anti-union labour relations. In the latter industries — the electrical industry, for example — managers adopted either a reformist or a highly authoritarian stance and used all kinds of modernist "organization" ideologies. It is not obvious whether this frame can be applied either to liberal countries like France or to an authoritarian country such as the USSR. Nevertheless, and leaving on one side the discussion of technological determinism, I wish to discuss how Maier's statements might still contribute to a more precise analysis of the relationship between production and politics.

The most radical industrial movement in the Soviet Union during the thirties was Stakhanovism. Dating from 1935, the movement was introduced as part of a much broader attempt to increase productivity. It was initially conceived by local party organizations that wanted to secure, in the mines of Donbass, a Union-wide achievement in the mining of coal. The principle was that people and resources would be placed at the disposal of a single productive shock

---

[1] Charles S. Maier, *In Search of Stability* (Cambridge: Cambridge University Press, 1988), p. 61.

worker, whose performance would be recorded. It was a very simple initiative and it commended itself immediately to the heads of the People's Commissariat of Heavy Industry: as was clearly perceived, in the drive for productivity, the initiative meant that the centre could rely upon the enthusiasm of well paid workers, even if the leaders of industry remained hesitant.[2]

It rapidly transpired that real and lasting improvements in productivity, won in accordance with the principles of the Stakhanovite movement, necessitated a profound reorganization that would impinge not only on the workplace but also on the whole production process. Production managers at every level of the hierarchical ladder began to discuss the means of implementing Stakhanovism. As they did so, some of them also raised the question of its value, so exposing themselves to the charge of wrecking the initiative. The dangers of this questioning were all too evident, since Stakhanovism had quickly assumed the character of a political movement dominated by the most radical sections of the Stalinist leadership. It provided these sections with an opportunity of targeting managers who, even before the introduction of Stakhanovism, had already been pursuing policies that were intended to enhance the profitability of industrial plants and to lead to greater managerial autonomy. This "struggle for profitability" was mostly conducted by industrialists in the metallurgical sector, supported by the head of the People's Commissariat of Heavy Industry.[3] Through Stakhanovism, the Stalinist general staff hoped to secure support in the most labour-intensive industries, and thereby to resist the influence not only of those it saw as "wreckers" but also of the new modernist industrial bureaucracy that emerged with the implementation of the successive five-year plans.

The moves that I have described reflect the state of industry in the USSR. What was being implemented, in fact, was a programme of economic reform that sought to carry through, in just a few decades, the process of industrialization that had already occurred, over a much longer period, in many European countries and the United States. But in the USSR it is the dominance of the primary sector in economic life which remains the main feature of the thirties (as of subsequent decades as well). This is illustrated not only by the initial years of planning and the launch of the Stakhanovite movement, but also by the introduction of forced labour, which began to be systematically organized in the

---

[2] The three main publications on Stakhanovism are Lewis H. Siegelbaum, *Stakhanovism and the Politics of Productivity in the USSR, 1935–1941* (Cambridge: Cambridge University Press, 1988); Francesco Benvenuti, *Fuoco sui sabotatori! Stachanovismo e organizzazione industriale in URSS: 1934–1938* (Roma: Valerio Levi, 1988); Robert Maier, *Die Stachanov-Bewegung 1935–1938. Den Stachanovismus als tragendes und verschärfendes Moment der Stalinisierung der sowjetischen Gesellschaft* (Stuttgart: Franz Steiner, 1990).

[3] Francesco Benvenuti, *Stakhanovism and Stalinism, 1934–1938* [CREES Discussion Papers, Soviet Industrialization Project Series, no. 30, July 1989] (Birmingham: University of Birmingham, 1989), p. 27.

Gulag in 1929, at about the time when the system of five-year plans was introduced.[4] The main architect of the system of forced labour was Nephtall Frenkel, who had been a very important international tradesman in wood before the war — a traditional and typically Russian profession. After 1917, he had probably become a GPU agent, before being arrested and sent to the Solovki, the first Soviet concentration camp. There he had investigated the best way of using the camp's workforce and had then been released at the start of the construction of the Belomor–Baltic canal, of which he was nominated a head.[5]

As early as the thirties, the Soviet Union had already been left behind in the sectors that were performing most dynamically in the economies of the West: the chemical and electrical industries. The contrast with France was stark. There, the three highest company turnovers in 1938, the date of the renewed take-off of French industry, were those of chemical and electro-metallurgical firms.[6] Soviet industrialization, on the other hand, was extensive, based as it was on huge plants with enormous workforces and on the tentative implementation of mass-production methods. Stress was placed much more on the intensification of work than on mechanization and automation — despite slogans such as the one that Stalin used in November 1935, when he called upon the Stakhanovites "to extract the maximum from technology".[7] The Soviet Union of the thirties believed that its industrialization was founded on the most modern American methods, but in reality it developed an industry of low technological intensity, with little being spent on research and development either for product or for process innovation. The chemical industry, the most technology-intensive industry, was a case in point. It ranked third, after metallurgy and oil, in the hierarchy of investments in the first five-year plan because it was based on advanced processes imported from abroad. Subsequently, however, no particular effort was made to improve it, and it remained neglected until the late fifties.[8]

As I have already indicated, the main driving force behind the Stakhanovite movement rapidly became a political one. It must be said that during the Stalinist period, and later, every major economic decision was in part, if not

[4] Oleg Vital'evich Khlevniuk, "Prinuditel'nyi trud v ekonomike SSSR 1929–1941 gg.", *Svobodnaia Mysl'* (1992), pp. 73–84.

[5] Alexander Solzhenitsyn, *L'Archipel du Goulag* (Paris: Seuil, 1972), vol. 2, pp. 61–3.

[6] Maurice Lévy-Leboyer, "La grande entreprise: un modèle français?", in Maurice Lévy-Leboyer and Jean-Claude Casanova (eds.), *Entre l'État et le marché. L'économie française des années 1880 à nos jours* (Paris: Gallimard, 1991), p. 377. See also Robert Boyer, "Le particularisme français revisité. La crise des années trente à la lumiére de recherches récentes", *Le Mouvement social* (1991), **154**: 3–40, on p. 18.

[7] Quoted in Siegelbaum, *Stakhanovism and Stalinism*, p. 97.

[8] Alcan Hirsch, *Industrialized Russia* (New York: The Chemical Catalog Company, 1934), pp. 58–61; R. Amann, M.-J. Berry, and R.W. Davies, "La science et l'industrie en URSS", in Eugène Zaleski *et al., La Politique de la science en URSS* (Paris: OECD, 1969), p. 449.

primarily, political. In the case of the Stakhanovite movement, and more widely during the thirties, what was at stake was the confidence which the political centre felt in industry as part of the defence structure of the USSR. This served as a powerful argument with which to persuade production managers not just to foster the movement but to take a leading role in it. As Ordzhonikidze, the People's Commissar of Heavy Industry and a very visible public partisan of Stakhanovism (in some ways even its creator), claimed, "There is no doubt that the Stakhanov movement will strengthen our country to such an extent that it will make [the USSR] so powerful, that no Hitler, no Japanese imperialist will ever dream of conquering a single patch of our Soviet land. They would never succeed in that!".[9]

Politics was in the ascendant, and the state authorities sought the politicization of all social relations. Andrei Andreev, in charge of the Central Committee's industrial department, suggested that the performance of industry, however good it might be, could never usurp the vital role of "politics", "vigilance", and the "class struggle" in the anticipated final victory of socialism.[10] Politically, the thirties became progressively tenser for production managers, even those who were trained during the Soviet period. Any departure from the plan could lead to judicial indictment. Joseph Berliner, who interviewed a number of pre-war managerial officers, reports that it was impossible to talk to any of them "without encountering some reference to imprisonment, sabotage, the secret police, informers, and so forth".[11] In fact, managers faced regular waves of repression between 1930, when those of them who had been trained during the tsarist period were targeted as "bourgeois cadres" and systematically removed from their posts, and 1937–8, when any of them might be prosecuted, arrested, sent to the Gulag, or simply shot in an NKVD cellar. Little by little, and despite resistance from people as powerful as Ordzhonikidze, the secret police had insinuated itself into factories and the whole economic apparatus, so enabling the public prosecutor to supervise very effectively the everyday activities of managers.[12]

All the industrial authorities and production managers, however, even those who became targets in the purges, were also politicians. For them, to produce pig-iron or to manufacture tractors was a political act, part of the fight for socialism; this was true from the factory engineers right up to Ordzhonikidze, who was a close friend of Stalin and who tried to protect the leaders of industry

---

[9] *Pravda*, 18 November 1935, quoted in Benvenuti, *Stakhanovism and Stalinism*, pp. 36–7.

[10] Benvenuti, *Stakhanovism and Stalinism*, p. 35.

[11] Joseph S. Berliner, *Soviet Industry from Stalin to Gorbachev. Essays on Management and Innovation* (Ithaca, N.Y.: Cornell University Press, 1988), p. 293, note 8.

[12] See, for example, Stalin's letter to Molotov, the President of the Council of People's Commissars, 12 September 1933, in Oleg Khlevniuk and Aleksandr Kvashonkin, *Politbiuro v 30-e gody*, vol. 1 of Andrea Graziosi (ed.), *Dokumenty sovetskoi istorii* (Moscow: Moscow State University, 1995).

from indiscriminate repression, until he committed suicide in February 1937. Nevertheless, there were things that Stalin's closest allies could not bear to hear. Typical was the following statement by Ordzhonikidze's deputy, Piatakov. In 1936 Piatakov wrote: "You cannot achieve anything with propaganda and agitation alone. This [the establishment of a leading role for the Stakhanovite movement] is a task which must be organized from an economic standpoint, and it must be well understood and supported from a technical standpoint".[13] Such a call for technological rationality could not be regarded by the Stalinist group as anything but reactionary.[14]

It may be asserted, therefore, that a clear watershed separates those who sought some kind of logic in management from those who were guided by political frenzy. Typically, the latter were most easily found among the managers of mines, a traditional source of support for Stalin since the Civil War.

## COMPANY OR FACTORY

I turn now to my second point. The political aims of every industrial manager in the Soviet Union were enshrined in the plan. The main political objective was to meet the plan's quantitative objectives and to do so in the right way. Yet the fulfilment of the Union-wide plan rested entirely upon the ability of the plant directors to meet their own more limited objectives.

Formed in 1932 from the Supreme Council of National Economy, the People's Commissariat of Heavy Industry was the most important economic ministry responsible for the acceleration of industrialization. Under its jurisdiction, there was a huge management apparatus, founded on specialized unions, each of which was responsible for the factories of a particular sector of industry. It was divided, in 1938, into several specialized Commissariats. Within this structure, the factory was the smallest unit: it was responsible for actual production and dependent on its union, the *glavk*. Although, formally, plant directors enjoyed no specific authority, in reality they wielded considerable power.[15] They had constantly to take personal initiatives in order to provide the factory with the necessary materials or machinery, which were in chronically short supply. To achieve this, they were obliged to deal unofficially, and often illegally, with other factory directors, specialized departmental heads, or even different industries. They had no choice but to determine by themselves the size and qualifications of the workforce which they needed. They

---

[13] *Pravda*, 30 June 1936, quoted in Benvenuti, *Stakhanovism and Stalinism*, p. 47.

[14] Piatakov was arrested in October 1936 and sentenced to death in the second Moscow trial in January 1937.

[15] David Granick, *Le Chef d'entreprise soviétique* (Paris: Editions de l'Entreprise Moderne, 1963), p. 75; translation of *The Red Executive. A Study of the Organization Man in Russian Industry* (New York: Doubleday, 1961).

also tried to put in place the most appropriate systems of wages and methods of organizing work, often in ways that conflicted with the national drive for rationalization. And at the most basic level of human existence, they had responsibility for the daily lives of their employees, providing food, housing, healthcare, and leisure and vacation facilities, among much else.

The heads of factories often found themselves at odds with the ministerial bureaucracy. They might, for example, oppose the action of the industrial unions that were in charge in their particular factory; their aim in this would typically be to prevent the unions from encroaching upon the production process.[16] Another very widespread practice consisted in developing, within their factory, the manufacture of those items which it was most difficult to obtain from other firms, such as rough castings, intermediary parts, spare parts, and, above all, machine tools.[17] In this way, from the time of the first five-year plan, it was neither the industrial union nor the ministry, but the factory that formed the true power unit within the centrally planned system — a situation that had not been foreseen by the state or by party legislation. It is as if some relay was needed inside the bureaucratic apparatus, in order to get the plans fulfilled, and as if the right level was the production unit.

This was an additional reason why many production managers were singled out as victims during the great purges. Because they wanted "to get things done" (to use an expression of Schumpeter's[18]), because they had a technical as well as a political passion for production, and because they were obliged to make their factories an actual unit of power, managers were vulnerable to the charge that they were seeking to build fiefdoms. Several were dismissed, arrested, and even shot on these grounds. In fact, the charge was not always false. But, as Francesco Benvenuti puts it, there was a surprising "growing official worry that industrial development may escape central control".[19] In reality, though, the factory proved itself to be the only stable structure in the Soviet economy.[20]

Obviously, business managers in the West did not have to fight for power in this way. There, the company rested on the contribution of an entrepreneur, who replaced the complicated structure of the market.[21] The Soviet production

---

[16] David R. Shearer, "Factories within factories: changes in the structure of work and management in Soviet machine building factories, 1926–1934", in William G. Rosenberg and Lewis H. Siegelbaum (eds.), *Social Dimensions of Soviet Industrialization* (Bloomington, Ind.: Indiana University Press, 1993), pp. 193–222.

[17] Granick, *Le Chef d'entreprise*, p. 76; Berliner, *Soviet Industry*, p. 282.

[18] Joseph Schumpeter, *The Theory of Economic Development*, trans. Redvers and Opie (Oxford: Oxford University Press, 1961), p. 93, quoted in Maier, *In Search of Stability*, p. 54.

[19] Benvenuti, *Stakhanovism and Stalinism*, p. 51.

[20] Granick, *Le Chef d'entreprise*, p. 77.

[21] R.H. Coase, "The nature of the firm", *Economica*, November 1937, pp. 386–405.

manager, on the other hand, powerful though he succeeded in being, did not run a company; he ran a plant, what can be considered as an administrative part of the state, an economic part of the collective capital.

## THE POLITICS OF THE PRODUCERS

Although French entrepreneurs were free enough within their territory, they also protected themselves constantly against the state or, alternatively, tried to exercise influence over it. Their objectives were numerous: they included a desire for protection against the labour laws,[22] the promotion of import regulations, and the struggle against business taxation. (On this last point, it is worth noting that the measures taken by the state to protect backward economic sectors often led to the overtaxing of large companies or to the neutralization of their action in competitive markets.[23]) The first social laws that began to limit the freedom of employers were those of 1928, on social insurance (*assurances sociales*), and 1932, on family allowances (*allocations familiales*). They were the first pieces of legislation to impose on French business the principle of a social obligation (a principle that had existed in Germany since the nineteenth century).[25]

Although French entrepreneurs exploited every means of political intervention, the period was not favourable towards large-scale industry or the urban centres in which most industrial activity tended to be located. As Richard Vinen has put it: "The political basis of the Third Republic lay outside the industrial regions in small towns and in the countryside".[26] As a way of coping with their problems, entrepreneurs might enter politics themselves. An example in the 1920s was Louis Loucheur, who was a patient advocate of industrial and social modernization and the only industrialist who held ministerial posts on several occasions between 1916 and 1930.[27] Other examples were the famous François de Wendel, who was a deputy or senator throughout the period from 1913 to

---

[22] Richard F. Kuisel, *Le Capitalisme et l'Etat en France. Modernisation et dirigisme au XXᵉ siècle* (Paris: Gallimard, 1984), p. 217.

[23] Lévy-Leboyer, "La grande entreprise", p. 406.

[24] Edouard Dolléans and Gérard Dehove, *Histoire du travail en France. Mouvement ouvrier et législation sociale*, vol. 2: *De 1919 à nos jours* (Paris: Domat Montchrestien, 1955), p. 20ff.; Alan Mitchell, *The Divided Path. The German Influence on Social Reform in France After 1870* (Chapel Hill and London: University of North Carolina Press, 1991).

[25] Maier, *In Search of Stability*, p. 53.

[26] Richard Vinen, *The Politics of French Business, 1936–1945* (Cambridge: Cambridge University Press, 1991), p. 23.

[27] Stephen D. Carls, *Louis Loucheur and the Shaping of Modern France, 1916–1931* (Baton Rouge and London: Louisiana State University, 1993).

1940,[28] and François Peugeot, who was elected to Parliament in 1936. Others again exploited powerful lobbying organizations, such as the Comité des Forges or the Comité des Houillères, with the intention of influencing governmental policy. After the victory of the Popular Front in 1936, a considerable number of employers' organizations developed. In the context of these, some of the most prominent personalities in the French industrial world, coming mainly either from the iron and steel industry or from certain more recently established sectors, such as the automobile industry, supported movements on the extreme right, either openly or surreptitiously. Conversely, one can find certain leaders of very modern industries, such as electricity, supporting *Les nouveaux cahiers*, which from its foundation in 1937 displayed a liberal approach to labour relations, while others in the same industries supported the Comité de Prévoyance et d'Action Sociale, an organization on the extreme right founded in July 1936.[29]

Despite this diversity, the most widespread political move in France on the part of industrialists and their senior staff between the first and the second world war was unmistakably the struggle against trade unionism, socialism, and communism. Broadly speaking, the managers of factories believed, or wanted to believe, that these movements were entirely an intrusion from outside. The firm had to be purged; otherwise the trade unions would remain as enemies within. An enduring French peculiarity in this area lies in the persistent lack of any structuring of industrial relations. In June 1936, however, the Matignon accords accepted the principle of collective agreements and the recognition of unions and their shop stewards inside companies; at the same time, the government granted, by law, the forty-hour week and paid holidays. But this was followed by two and a half years of open struggle against trade unionism and communism within firms.[30] In this period, the French *patronat* brought decisive pressure to bear on successive governments in an attempt to counter labour laws, especially the one concerning the forty-hour week.[31] Finally, these efforts had their effect on government policy, when in 1938 the *patronat* secured its long-awaited revenge with the end of the Popular Front and the advent of a resolutely liberal government.

---

[28] Jean-Noël Jeanneney, *François de Wendel en République. L'argent et le pouvoir, 1914–1940* (Paris: Seuil, 1976).

[29] Vinen, *The Politics of French Business*, p. 60ff; Patrick Fridenson, *Histoire des usines Renault*. vol. 1: *Naissance de la grande entreprise, 1898/1939* (Paris: Seuil, 1972), Appendix.

[30] See Adrian Rossiter, "Experiments with corporatist politics in Republican France, 1916–1939" (University of Oxford D.Phil. thesis, 1986), chapter 5.

[31] See Ingo Kolboom, *La Revanche des patrons. Le patronat français face au Front populaire* (Paris: Flammarion, 1986); originally published as *Frankreichs Unternehmer in der Periode der Volksfront. 1936–1937* (Rheinfelden: Schäuble Verlag, 1983).

The main economic issue of the thirties was the continuing depression, one of the characteristics of which was the persistently low level of investment. All parties shared the goal of diminishing investment as a way of sharing work between as many workers as possible and reducing overproduction.[32] It is striking that, from the beginning of the Great Depression in France, i.e. from 1931, until 1936, even those industrialists who were most dependent upon the world market were at one with successive governments in supporting a deflationary policy. The policy had succeeded during the late twenties, and nobody wanted to break with it now, although it had the effect of maintaining the country in a state of depression.[33] It was only with the advent of the government of the Popular Front that the policy was finally abandoned. A main architect of this plan, which replaced the policy of deflation with one founded on devaluation, was Léon Blum, the socialist prime minister. Obliged to intervene in the economy far more than he had expected, Blum was unable to implement the traditional policy of the state towards industry. The state, as he saw it, had to display determined support for the modernization of the mechanisms of production. The consequence was a massive investment in the nationalized companies, mainly in aeronautics — a choice that already intimated the prospect of rearmament. The change of direction was decisive. From 1938, after another, very brief Blum cabinet, the Daladier government assumed responsibility for an economy in which, over two years, the idea of an "économie dirigée" had made irreversible progress. Thereafter, the economic policy of the state displayed greater consistency, with armament production, for example, being systematically rationalized.[34]

In the Soviet Union, production managers and industrial specialists had some success, unexpectedly, in their efforts to influence Politburo policy. This was the case between 1930 and 1933, when they intervened effectively in the dramatic issue of planning realism. The ever higher targets of successive annual plans during the period of the first five-year plan provoked considerable resistance among high-ranking industrial leaders. The dramatic failure to reach the goals of increasing production by 45 per cent in 1931 and of achieving 17 million tons of pig-iron in 1932 (only 6.2 millions were actually obtained) had been foreseen by specialists. It led to the revision of the targets of the second five-year plan, which was to begin in 1933. Many managers and specialists criticized more or less openly the whole process of planning. In R.W. Davies's words, "while all these complaints were being voiced, the Politburo gradually and painfully adjusted the plans downward toward reality". In January 1933, Stalin announced "an average annual increase in industrial production of 13–14 per cent as a

---

[32] Lévy-Leboyer, "La grande entreprise", p. 392.

[33] Boyer, "Le particularisme", p. 10.

[34] Kuisel, *Le Capitalisme*, p. 211; André Straus, "Vers une restauration de l'Etat", in Lévy-Leboyer and Casanova, *Entre l'État et le marché*, pp. 274–87.

minimum" for the second five-year plan — a figure significantly lower than the 22 per cent to which the first plan aspired.[35]

It was so easy to be accused of wrecking that it required great political courage to oppose Politburo decisions and targets enshrined in the plan and to propose what seemed to be more "realistic" measures. A whole group of railway special-ists was sacked in 1935 for claiming that "railways could in no way improve their performance unless substantial investments were made". This claim was termed a typical example of "anti-Soviet limit theory".[36] As often happened, a point of view based on techno-economic arguments had been judged to be political or, more precisely, "reactionary" by the small group of radical Stalinists, in this case by Lazar Kaganovich, who had been in charge of communications since February 1935. As a matter of fact, when Soviet industrial managers and specialists adopted technoeconomic stances of this kind — Piatakov's stance on Stakhanovism is an example — they were in no sense opposing Soviet power: their response was, on the contrary, an attempt to protect the Soviet Union against itself and to have themselves numbered among its "best sons". As a result, the crucial challenge to factory managers was to make the right choice of allies within the hierarchy, up to the very top of the relevant departments or People's Commissariats and of the Central Committee.

To sum up, therefore, the policies of the production managers extended into political action in both countries, and a political judgment became a component of production activity both within and outside the factories.

## POLITICAL SYSTEM, PRODUCTION, AND INNOVATION

Finally, I turn to some suggestions concerning the relations between political systems and innovation.

In the summer of 1932, Stalin wrote: "We have a passion for construction, and that's excellent, but we lack the passion for mastering production". Soon afterwards, mastering was changed to assimilation, and the assimilation of technologies and modes of production became the main objective of the second five-year plan, giving a pretext for "the switch to a slower pace of industrialization".[37] But whether Stalin talked of mastering or assimilation, he was speaking about himself. It was he, as the first leader, and the "boss" (*khoziain*) of the country under one-party rule, who had to recognize the need to master and assimilate production and technology. In the planned Soviet system, such discourse

---

[35] R.W. Davies, "The management of Soviet industry", in Rosenberg and Siegelbaum, *Social Dimensions*, pp. 104–123, on p. 110f. (for the whole paragraph).

[36] Benvenuti, *Stakhanovism and Stalinism*, p. 19.

[37] Davies, "The management of Soviet industry", p. 114.

did not help to provide factory managers with better conditions for innovation. The latter were under the thumb of the "ratchet principle", which means, as Joseph Berliner explains, that "once a new high level of performance has been achieved, the next plan target may not be reduced below that level"; the target had usually to be raised still further.[38] This principle does not foster innovation in product or process. From the broader perspective of the Soviet iron and steel industry, a five-year period, in which adjustments were introduced by successive annual plans, was not too damaging for the implementation of innovation. But it was very difficult to give any leeway that would allow new technological domains to develop and gain prominence. Chemistry is a good example here, as electronics and data processing were to be, cruelly, after the war.[39]

The innovation process was itself highly centralized. Every industrial commissariat and, within each of these, every main production department had a technical department or section in charge of R and D. In this structure, design was separated from production, even though some large factories had their own research departments. For most products, the decisive choice was made by the Politburo itself. This was also, of course, the case for all military material (although for this, as a priority product, more flexible planning controls were imposed).[40]

Basically, the Soviet economy was not consumption-oriented, but politics-oriented. Whereas western entrepreneurs formed an integral part of economic life, and whereas Louis Renault could claim in 1932 that "To live is to consume",[41] the Stalinist production manager could do no more than implement, as best he could, an economic policy fashioned in the political sphere. There were rigorous constraints on the extent to which the Soviet factory director could spread his "joy of creating, of getting things done" (to use the phrase that Schumpeter coined with regard to entrepreneurs). In contrast, French entrepreneurs could decide, in 1935, to merge the design and the production of a small popular car, and then, two years later, to abandon the project.[42] They could even display political imagination, as happened in 1936, when Renault, a

---

[38] Joseph S. Berliner, *Factory and Manager in the USSR* (Cambridge: Harvard University Press, 1957), p. 78.

[39] Yves Logé, *URSS. Le Défi technologique. La révolution inachevée* (Paris: Presses Universitaires de France, 1991), p. 49.

[40] See Kendall E. Bailes, *Technology and Society under Lenin and Stalin* (Princeton, NJ: Princeton University Press, 1978).

[41] Fridenson, "L'idéologie des grands constructeurs dans l'entre-deux-guerres", *Le Mouvement social* (1972), **81**: 51–68, on p. 51.

[42] Patrick Fridenson, "La question de la voiture populaire en France de 1930 à 1950", *Culture technique* (1989), **19**: 205–10.

leading figure in the electrical industry, and a number of bankers proposed the creation of a United States of Europe, based on an alliance between France and Germany.[43]

In conclusion, the contrast between the two countries I have discussed can be summarized succinctly. The Soviet Union trained a production manager-politician with no concern for the market and with a reluctance to innovate. In France, by contrast, it was the market that pushed entrepreneurs into politics, through a pressure for technical as well as for political innovation.

---

[43] Fridenson, "L'idéologie des grands constructeurs", p. 56.

# Managing Complexity: Interdisciplinary Advisory Committees

*Thomas P. Hughes*

## MESSY COMPLEXITY

A characteristic of engineering and applied science in the United States and other highly industrialized nations is their enlarging capacity to build increasingly complex weapons, production, energy, transportation, communication, chemical, and information systems. Individual creativity may not have increased demonstrably in recent centuries, but groups and networks of humans today can create and construct on a far larger scale and far more complexly than in the past. Especially since the second world war, engineers, scientists, and managers have increasingly, perhaps unwittingly, opened the Pandora's box of technological complexity, and, as a result, have had to respond to a "crisis of control".[1]

Contending with complexity is not limited to those who try to manage large technological systems. Architects, sociologists, literary critics, and many others see complexity as a defining characteristic of a postmodern industrial world. In 1966, Robert Venturi, an architect, wrote:

I like complexity and contradiction in architecture.... I speak of a complex and contradictory architecture based on the richness and ambiguity of modern experience, including that experience which is inherent in art. Everywhere, except in architecture, complexity and contradiction have been acknowledged, from Gödel's proof of

---

[1] On crises of control, see James R. Beniger, *The Control Revolution. Technological and Economic Origins of the Information Society* (Cambridge, Mass.: Harvard University Press, 1986).

ultimate inconsistency in mathematics to T.S. Eliot's analysis of 'difficult' poetry and Joseph Albers's definition of the paradoxical quality of painting.

Denise Scott Brown, Venturi's partner, has written:

> ... design is the subtle organization of complexity, the orchestration of sometimes inharmonious instruments, the awareness that discord at a certain level can be resolved as harmony at another. It requires patience. It is a pinpoint upon which it is difficult for the professional to live.[2]

Sociologists David Harvey and Zygmunt Bauman articulate a postmodern language, the common denominator of which is complexity. They refer to indeterminacy, heterogeneity, fragmentation, pluralism, contingency, ambivalence, non-linearity, and distrust of totalizing discourses. Indeterminacy finds expression in catastrophe and chaos theory as well as in fractal geometry. In the postmodern discourse, "complex system" replaces "system" in its earlier modern, orthodox, and organismic sense.[3] The vocabulary of postmodernity is poles apart from that of modernity.

Ecologists, molecular biologists, computer scientists, systems analysts, and organizational theorists have also familiarized us with concepts of complexity. In the United States, the military has funded a number of scholars intent on exploring complex systems, among them John von Neumann, George Dantzig, Philip Morse, and Jay Forrester. Nobel laureates Herbert Simon and Ilya Prigogine have thought and written about complex systems, as have Friedrich von Hayek and Ludwig von Bertalanffy. The list can be greatly extended.[4]

There are countless definitions of complex systems. Among the most succinct and lucid is that of Joel Moses, mathematician, computer scientist, and provost at the Massachusetts Institute of Technology, who defines a complex system as one with many non-simple, non pattern-repeating connexions. He does not equate complexity with the number of functions the system is able to perform.[5] The heterogeneity, as contrasted to the homogeneity, of the components of a system contributes to its complexity, assuming that this heterogeneity results in the increase of the heterogeneity of the connexions. When patterns can be found in the component and connexion structure, complexity can be reduced. This definition is abstract enough to cover both technical and organic systems.

---

[2] Denise Scott Brown, "Between three schools", *Architectural Design* (1990), **60**: 8–20.

[3] I have drawn the vocabulary of the modern and postmodern from the introduction and chapter nine of Zygmunt Bauman, *Intimations of Postmodernity* (London: Routledge, 1992). See also David Harvey, *The Condition of Postmodernity. An Enquiry into the Origins of Cultural Change* (Oxford: Basil Blackwell, 1989), p. 9.

[4] David Warsh, *The Idea of Complexity* (New York: Viking, 1984), pp. 3–4.

[5] Joel Moses, "Organization and ideology: paradigm shifts in systems and design", unpublished monograph (Cambridge, Mass.: MIT, 1991), pp. 114–21.

## MANAGERIAL RESPONSES TO COMPLEXITY

In this century, system builders, individual and collective, have constructed sociotechnical or technological systems of messy, nearly overwhelming complexity. Because of the construction of massive technological systems since 1940 in the United States, a revolution in managerial techniques comparable to that associated with the introduction of Taylorist scientific management around 1900 has occurred. The military, especially the Air Force and the Navy, funded many of the persons, organizations, and programmes that helped to usher in the new management style. Academic and industrial engineers, scientists, and managers, allied with a small, innovative group of military officers, led this managerial and conceptual revolution.

These new techniques of management have been applied to both operating systems and systems under construction. Alfred D. Chandler, a distinguished historian of modern management, argued years ago that managerial innovations were more responsible for the second industrial revolution than technical ones. Chandler provided a detailed and conceptually rich account of the the new management of modern multi-divisional manufacturing firms during the period from about 1870 to 1950.[6] He has not, however, dedicated his keen analytical and synthetical approach to describing and explaining the messy, heterogeneous, multi-faceted managerial style used by system builders (programme managers and project managers) engaged with the construction of research and development-intensive, military-funded construction systems that flourished during and after the second world war.[7] I am persuaded that their managerial innovations made possible many of the leading technological advances of recent decades.

Major managerial innovations arising from this system-building activity include the introduction of programme and project management, systems engineering, and systems analysis; the use of computer-based communication, control, and command networks; and the resort to interdisciplinary advisory committees. In this essay, I focus on the interdisciplinary advisory committees of scientists, engineers, and managers that have been relied upon by system builders constructing military-funded systems. (In forthcoming essays, I shall consider other managerial innovations.) The committees considered here helped conceptualize research and development projects and organizational structures to preside over them. Such collective decision-making about research and development sharply contrasts with decision-making by individuals in earlier eras.

The managerial innovations I discuss have long and complex antecedents, but I start my account of interdisciplinary committees with their use during the

---

[6] Alfred D. Chandler Jr., *Scale and Scope. The Dynamics of Industrial Capitalism* (Cambridge, Mass.: Belknap Press of Harvard University Press, 1990).

[7] Stephen P. Waring provides a recent historical account of managerial innovations in *Taylorism Transformed. Scientific Management Theory Since 1945* (Chapel Hill, N.C.: University of North Carolina Press, 1991).

wartime Manhattan Project, or, more precisely, the project managed by the Manhattan Engineering District (MED) of the U.S. Army Corps of Engineers, a project that is a watershed in the history of large technological systems. After MED, I concentrate on the 1950s and describe the use of interdisciplinary advisory committees in connexion with the design, research, and development of systems to manufacture and assemble the components of the SAGE air defence system. So, here I focus on the creation of weapons systems, not on the systems after they were deployed.

## MANHATTAN ENGINEERING DISTRICT: A WATERSHED

The history of large, complex technological systems entered a new stage during the second world war with the activities of the Manhattan Engineering District. The MED project became the paradigm, the inspiring myth guiding postwar builders of military-funded systems. Today the project would be characterized as a military-funded programme involving a number of subsystems, or projects, with the MED organization acting as programme manager. In essence, MED created a production system for making explosives and assembling bombs.

General Leslie Groves, Army Quartermaster and Corps of Engineers veteran, who had been in charge of the construction of the Pentagon building and had demonstrated outstanding abilities on other large construction projects, was in command of MED. Groves had studied engineering at MIT for two years before going to West Point, from where he graduated fourth in his class in 1918. Working with Groves and formally categorized as serving under his command in the Army's hierarchical structure were a number of organizations that served as project managers, as contrasted to the programme manager with its broad overarching responsibilities that subsumed a number of projects. Best known among the project managers was the Los Alamos Laboratory, headed by Robert Oppenheimer. The Los Alamos laboratory presided over the prototype design, development, testing, and assembly of the bomb. Explosives for the bomb came from Hanford and Oak Ridge. Other project managers included Stone & Webster Engineering Corporation, which designed and constructed the electromagnetic separation project for producing uranium 235 at Oak Ridge, Tennessee; and the Du Pont Company, which had charge of the plutonium reactor project at Hanford, Washington. These project managers co-ordinated the activities of a number of subcontractors, including general contracting companies and component manufacturers.[8]

---

[8] This account of MED differs in detail from my account given in Thomas P. Hughes, *American Genesis. A Century of Invention and Technological Enthusiasm 1870–1970* (New York: Viking, 1989), pp. 381–421, primarily because the point of view taken in this essay emphasizes the role of corporate programme and project management and of interdisciplinary committees.

MED, like similar postwar military-funded programmes, is notable as an organization-transcending enterprise. United States government contracts tied together a host of industrial firms and university laboratories into a military-industrial-university complex. As a contractual system, MED resembles earlier military-construction projects, including the building of the Pentagon, but university laboratories did not become major actors in these earlier projects, and the research and development activity was limited. Historians and others who have written about MED, however, have often gone astray in seeing the endeavor as primarily a scientific research and development programme. MED was essentially a construction programme involving research, design, and engineering that culminated in a production system. This is why the Army chose Groves to head the project.

Though it was a construction programme, much like the Pentagon programme, MED differed significantly because of the need for extensive research. This probably caught Groves unaware and unprepared. The role of the research scientists and engineers is similar in many ways to that of the Thomas Edison team during the invention and preliminary design of the electric light system. Before the Army took over management of the American nuclear bomb programme, academic scientists and engineers had generally enjoyed freedom of action comparable to that experienced by Edison and the other American independent inventors during their halcyon days late in the nineteenth century. Working in small, informal university laboratory environments, nuclear physicists, chemists, and metallurgists, like Edison and the independent inventors, were unencumbered by corporate interests and bureaucratic structures. After the Army took charge and brought in architect-engineering companies and industrial corporations, such as the Du Pont Corporation, Stone & Webster Engineering Corporation, and the Kellex Corporation, Groves and his staff pressured MED scientists at Columbia University, the University of Chicago, and the University of California (Berkeley) into playing roles not unlike those of scientists in an industrial research laboratory involved in the development of a new system of production. They conducted theoretical studies and experiments to obtain data needed by design and production engineers. This distressed MED scientists, especially Eugene Wigner and Leo Szilard at the University of Chicago Metallurgical Laboratory, because they were familiar with the European practice of appointing scientists to head programmes and projects.

## ADVISING MED

Interdisciplinary advisory committees that included engineers as well as scientists played a major role in the nuclear bomb programme during the early, preliminary design phase before and after the Army took charge. The Office of Scientific Research and Development (OSRD), established in June 1942 and headed by former MIT electrical engineer and vice-president, Vannevar Bush, had responsibility for contracting for early conceptual and background-research studies for the uranium project. Bush delegated this assignment in June 1942 to an OSRD executive committee with the code name S-1. James Conant, president of Harvard University, chaired the committee, which included eminent scientists and one engineer, Eger Murphree, vice-president of the Standard Oil Development Company, who were familiar with nuclear physics. They proposed research agendas and preliminary designs for several processes for the production of explosive fissionable material. By late 1942, after MED had become programme manager of the uranium project, the scientists and engineers decided on the gaseous diffusion and electromagnetic separation processes for making uranium 235 and the nuclear reactor process for making plutonium. Scientists Harold Urey of Columbia University, Ernest Lawrence of the University of California (Berkeley), and Arthur Compton of the University of Chicago, who headed the research endeavors for each of these projects, sat on the S-1 executive committee with Conant. A planning board, headed by Murphree and consisting of four other outstanding engineers, organized engineering studies and supervised pilot plant experiments preliminary to the production of atomic explosives.

Capable of organizing an industrial contractual system, Groves was not sufficiently competent in science and advanced engineering to make critical decisions alone about the details of design, research, and development. So constrained, he repeatedly turned to MIT engineering professor Warren K. Lewis, a founder of the field of chemical engineering, to establish inter-disciplinary advisory committees to assist in decision making. "Doc" Lewis, who began teaching at MIT after receiving his doctorate in chemistry in Germany before the first world war, had forged close relations between MIT and industry, especially the rapidly growing petroleum refiners. In 1916 Lewis and his MIT colleague, chemist William H. Walker, established the School of Chemical Engineering Practice that made possible the participation of MIT graduate students and professors in corporate research projects. Professors also served industry frequently as consultants, and the corporations hired MIT graduates.

On the undergraduate and graduate levels, Lewis trained a disproportionate number of the heads of the major petroleum refining and chemical companies to which he often acted as consultant. Owing in no small measure to his teaching and the rise of the petroleum industry, the United States by the onset of the second world war had large numbers of chemical engineers grounded in basic chemistry and physics and experienced in designing and managing large industrial projects. Lewis trained his students and encouraged the chemical and refining companies he advised to apply an overall systems approach to the design and construction of chemical and refining facilities.

Lewis served on a surprisingly large number of important advisory committees throughout the duration of the atomic bomb project. In October 1941, before the United States entered the conflict, he joined the scientists on a National Academy of Sciences Committee chaired by Compton. The committee recommended an all-out effort to design, develop, and produce atomic bombs.[9] Early in 1942 he joined the Planning Board, which included former Lewis students. Membership included L. Warrington Chub, director of the Westinghouse Research Laboratories; Percival C. Keith, who had studied under Lewis and then become head of the M.W. Kellogg Company; and George O. Curme, who repre-sented the Union Carbide & Carbon Company. Murphree, another of Lewis's former MIT students, chaired the group. Union Carbide & Carbon and M.W. Kellog Company subsidiaries soon became major MED contractors, the latter responsible for designing and constructing a gaseous diffusion device and the former for operating it.

After MED became programme manager, General Groves and Conant established a review committee headed by Lewis in November 1942 that made site visits and evaluated the major research projects under way at Columbia University, the University of Chicago, and the University of California, (Berkeley). All members of the committee had been Lewis students. The committee's report crystallized the decision to initiate not one but three parallel MED projects: the construction of plutonium-producing reactors and related plutonium-separation plants; the construction of electromagnetic separation facilities for making U-235; and the construction of a gaseous diffusion plant for producing U-235. In January 1943 Lewis led still another committee that evaluated plans for a thermal diffusion plant to make U-235: thermal diffusion became the fourth MED construction project. In July 1943 Lewis chaired yet another committee, this time to review plans of Chicago scientists to design and develop a heavy-water reactor as an alternative to the Du Pont plutonium reactor design. Lewis and this committee also heard the long-standing, even bitter, complaints of Chicago scientists about Du Pont's putative usurpation of the Chicago laboratories' responsibilities and dissatisfaction with the chemical company's policies as manager of the plutonium reactor project.[10]

Calling again on Lewis in March 1943, Groves appointed him to chair an interdisciplinary committee to evaluate the research plans being formulated at the Los Alamos Laboratory headed by Oppenheimer. Other members of this committee were Edwin Rose, an ordnance specialist; John H. Van Vleck, a theoretical physicist; and E. Bright Wilson Jr., a Harvard chemist and explosives expert. Physicist Richard Tolman served as secretary. Among other recommendations, the committee suggested that Los Alamos take on additional metallurgical responsibilities, including the purification of plutonium. The committee also called for the establishment of an engineering division at Los Alamos. As a

---

[9] Arthur Compton, *Atomic Quest* (New York: Oxford, 1956), pp. 49–50.

[10] Richard G. Hewlettt and Oscar E.Anderson, Jr., *A History of the United States Atomic Energy Commission. The New World, 1939/1946* (University Park, Pa.: Pennsylvania State University Press, 1962), pp. 203–4. I have relied on Hewlett and Anderson's account of Lewis' committee contributors. On the organisation of MED, see Vincent C. Jones, *Manhattan: The Army and the Atomic Bomb* (Washington, D.C.: Center of Military History, U.S. Army 1985), pp. 73–92.

result, the laboratory became more than the small physics laboratory originally conceived. Addition of the engineering component doubled the size of the staff.[11]

Reviewing Lewis's role as a senior adviser for MED, we see the importance of advisory committees for MED and the interdisciplinary character of several of these committees, as well as the important position of engineers on these committees. Because of the dramatic events at Oppenheimer's Los Alamos laboratory, which designed and tested the bomb, and because of the charismatic personalities of many of the physicists working there, historians and journalists have tended to overlook the role of engineers, skilled workers, and managers in MED. After the war, engineers soon took a similarly important role on advisory committees associated with postwar military-funded, large-scale programmes.

## PROJECT SAGE: ADSEC

Turning to the 1950s, I now focus on the Semi Automatic Ground Environment (SAGE) air-defence system, a research and development project of notable size and complexity that was a part of the armed services air defence programme. SAGE engendered several outstanding organizational innovations. Salient among them is the role taken by the MIT, an intricately articulated research university, as a principal system builder for this billion-dollar, military-university-industry programme. This unusual, perhaps unique, performance by a major university merits investigation and reflexion by historians as well as by policy makers. In addition to MIT as system builder, interdisciplinary committees of academic and industrial engineers and scientists, continuing in the tradition of Lewis committees, took the lead in defining SAGE research and development problems and in suggesting solutions. Because of university leadership, other new organizational forms, such as "summer studies"; a mission-oriented, university-managed research and development laboratory; computer-based command, control, and communication; and a non-profit, military-funded contract research and development centre mushroomed throughout the programme. In this essay, I concentrate on the history of the interdisciplinary committees.

SAGE history begins with the explosion of a fission bomb by the Soviet Union in 1949, an event that caused the administration of President Truman concern about the neglected state of its continental air defence system. The Defence Department placed the burden of responding to the problem upon the Air Force, which in turn ordered its Scientific Advisory Board (SAB), to investigate means of radically strengthening ground defence against long-range

---

[11] Hewlett and Anderson, *History of the United States Atomic Energy Commission*, pp. 235–6.

bombers carrying atom bombs. This committee, headed by Theodore von Kármán, a Hungarian emigré with an international reputation as an aeronautical engineer, and with members from academia and industry, had been set up to utilize the nation's engineering and scientific talent for the benefit of the Air Force. Having separated from the Army only two years before and unburdened by in-house research and development organs, the Air Force turned outside its own ranks to establish a research and development support system.

Kármán's committee established a subcommittee, the Air Defense Systems Engineering Committee (ADSEC), which was to play a key role in formulating design concepts for SAGE. The head of the new committee was George E. Valley, Jr., an associate professor of physics at MIT who had worked in the MIT Radiation Laboratory during the war, who had helped edit its seminal technical reports after the war, and who had been serving on the Electronics Panel of the SAB. He turned for assistance to John Marchetti, civilian director of the Air Force Cambridge Research Laboratories (AFCRC), who, along with his staff of veterans of the MIT Radiation Laboratory and the wartime Harvard Radar Research Laboratory, had studied problems of improving the use of radar in air defence.[12]

Valley chose an interdisciplinary committee with academic members from engineering, electronics, and aerodynamics. Members of ADSEC came from the northeast section of the country, he argued, to facilitate easy consultation.[13] Valley suggested that another committee might be needed to respond to the situation on the West Coast. MIT faculty and researchers dominated the committee: Charles S. Draper, a SAB member, aeronautical engineer, and head of the MIT Instrumentation Laboratory; H. Guyford Stever, also a SAB member from MIT, an aeronautical engineer, and expert on guided missiles; and Henry Houghton, a SAB member and head of MIT's department of meteorology. Others included Marchetti; A. Donovan, a SAB member and an aerodynamicist from Cornell University; and G. Comstock from Airborne Instruments Laboratory, another radar expert. This committee took on the problem of conceiving a basic design for an improved air defence system and recommending a related research and development programme.

ADSEC began meeting in Cambridge, Massachusetts, in December 1949 and at the peak of its activities convened every Friday. It began by studying histories of the Royal Air Force air defence during the second world war and surveying existing air defences in the United States. These systems depended on humans processing information received by voice, telephone, communication from radar,

---

[12] George E. Valley, "How the SAGE development began", *Annals of the History of Computing* (July 1985), **7**: 198.

[13] Kent C. Redmond and Thomas M. Smith, *Project Whirlwind. The History of a Pioneer Computer* (Bedford, Mass.: Digital Equipment Corporation, 1980), p. 172.

and ground observers. Personnel plotted the tracks of attacking and defending aircraft on large display boards, familiar to many of us from action films. Voice radio directed interceptors to their targets. This was no automated system.[14]

Members of ADSEC conceived their task as one of "automating" the air defence system. Automation had become a trend, even trendy, because of Norbert Wiener's publication of *Cybernetics* (1948) and because of essays by John Diebold and other management consultants who lauded the formation in manufacturing companies, such as the Ford Motor Company, of automation departments staffed by automation engineers.[15] Other engineers had for decades been developing servomechanism and feedback controls, a case in point being the inventions of Elmer Sperry, and the work of the MIT Servomechanisms Laboratory, established by the MIT electrical engineer Gordon Brown on the eve of the second world war.

ADSEC initially found the existing non-automated air defence "depressing but not discouraging"; but soon its judgment became harsher.[16] Not discouraged but forward looking, the Valley committee proposed replacing the few large radars with several thousand mostly unattended serially produced small radars able to track both low- and high-altitude aircraft. The committee believed that ordinary telephone lines could be used to transmit coded information from the radars to "data analyzers" which in turn would be linked together to an area command centre by telephone lines. The reference to data analyzers anticipated the use of computers, but in its early deliberations the committee had only the haziest notion of what an analyzer would be like. [17]

ADSEC not only saw itself as introducing automation to air defence, it also believed itself to be pioneering in conceptualizing air defence as a system. The committee's effort to define the term "system" gives us an idea of the state of systems thinking at the time. Rather than seeing this semiotic exploration as a philosophical diversion in a sober technical report, we should understand the search as an effort to conceptualize organizational means of systematizing the heterogeneous components that would constitute an air defence system. ADSEC declared:

> The word itself is very general ... [as in] the "solar system" and the "nervous system", also in which the word pertains to special arrangements of matter; there are also

---

[14] Charles J. Smith, *History of the Electronic Systems Division. SAGE. Background and Origins* [AFSC Historical Publication Series, 65-30-1] (Bedford, Mass.: Historical Division, Office of Information, Electronic Systems Division, 1964), vol. 1, pp. 68–9.

[15] James R. Bright, "The development of automation", in Melvin Kranzberg and Carroll Pursell (eds.), *Technology in Western Civilization* (New York: Oxford University Press, 1967), vol. 2, pp. 635–7

[16] Smith, *History of the Electronic Systems Division*, vol. 1, p. 69.

[17] ibid., vol. 1, p. 71.

systems of philosophy, systems for winning with horses, and political systems; there are the isolated systems of thermodynamics, the New York Central System, and the various zoological systems. [18]

The committee decided that a defence system resembled the category of system known as organisms. Resorting to an extended metaphor, as imaginatively inventive persons often do in moving from the known to the unknown, the committee likened radar and microphones to eyes and ears, data processors to human cognition, and telephone linkages to the body's information transmission network. Having done so, the committee, pursuing its anthropomorphic metaphors, concluded that the existing air defence system was "lame, purblind, and idiot-like".[19]

To give the purblind system sight, Valley organized Thursday evening radar seminars attended by mathematicians, engineers, and physicists from industry as well as from universities and Air Force installations. Among those participating in the long discussions were engineers and scientists from the Rand Corporation and from Raytheon Corporation's guided missile group. "All these people struggled to invent a new solution to the ground clutter problem". The seminars stimulated a number of activities, leading Valley to claim that "numerous inventions made in many laboratories and a huge number of journal articles have resulted from this work".[20] To make the existing system less "idiot-like", ADSEC recommended that the Whirlwind electronic digital computer, then being developed at MIT by Jay Forrester and his group, be incorporated into the system and that, in preparation for this, a small pilot system using this computer, several radar stations, and a number of aeroplanes be used for testing and experimentation of air defence. The Valley committee did not simply review research and development projects; it designed them.

## SAGE: PROJECT CHARLES

To do the research, to design the necessary hardware and computer programmes, and to develop a full-scale system, Valley and his associates also decided, after a year's deliberation, that a special-purpose laboratory was needed. At the suggestion of Valley, Air Force Chief of Staff General Hoyt S. Vandenberg asked MIT to establish an Air Force-funded laboratory. Memories of the Institute's

[18] Air Defense Systems Engineering Committee, "Air Defense System, Report of 24 October 1950", p. 2.

[19] Air Defense Systems Engineering Committee, "Air Defense System, Report of 24 October 1950", pp. 9–10.

[20] George E. Valley. "How the SAGE development began", *Annals of the History of Computing* (July 1985), **7**: 211.

success in establishing and administering the war-time Radiation Laboratory remained fresh in the minds of those recommending MIT. In a counterproposal, the chancellor of MIT, Julius Stratton, with the concurrence of MIT president James Killian and the advice of the scientific advisory committee of MIT's Research Laboratory of Electronics (a spin-off from the Radiation Laboratory), proposed that a more thorough study be carried out than that done by ADSEC and that all three branches of the military should take part in funding the new laboratory. Jerrold Zacharias, an influential physicist at MIT who advised Stratton and Killian, believed that ADSEC was recommending a technical, or quick, fix that did not take into account social, political, and economic factors, such as the dispersal of critical industry and the construction of air raid shelters.[21] In response to the Air Force request, Stratton proposed to the Air Force an extensive "summer study project" to make recommendations about the design of air defence as well as to reach a decision about MIT's role in the establishment of the proposed mission-oriented laboratory.[22]

Summer studies, an innovative organizational form that provided connective tissue between the military and academia, was becoming a hallmark of the MIT research and development style.[23] Following the suggestion of academics, the military used the programme to obtain the advice of academics during the months when they were usually free from teaching responsibilities. (In fact, some of the so-called summer study projects extended over a longer period.) In essence, the summer study groups were interdisciplinary, academic committees meeting continuously over an extended period rather than from time to time. Usually they met in some secluded, even bucolic, retreat-like setting.

Summer studies have generally been positively evaluated. In some cases, however, those participating concluded that the military sponsor intended scientists to legitimate decisions already made. Some scientists, Zacharias outstanding among them, also felt that the military tended to define the complex problems assigned to summer study groups too narrowly. He judged the approach taken by the designers of Project Lexington, the first MIT-sponsored summer study, too narrowly technical. Project Lexington, by focusing on the technical feasibility of atomic propulsion for aeroplanes, largely ignored costs and alternatives.[24]

Having designed and organized Project Hartwell, the second and outstandingly successful summer study that became a model for others, Zacharias became known for his insistence that the definition of the problem for a summer study be broad, so that the engineers, scientists, and other participants could take

---

[21] Jack S. Goldstein, *A Different Sort of Time. The Life of Jerrold R. Zacharias. Scientist, Engineer, Educator* (Cambridge, Mass.: MIT Press, 1992), pp. 113–14.

[22] James R. Killian, Jr., *The Education of a College President. A Memoir* (Cambridge, Mass.: MIT Press, 1985), pp. 71–2.

[23] J. R. Marvin and F. J. Weyl, "The summer study", *Naval Research Reviews* (August 1966), **20**:1–7ff.

[24] Goldstein, *A Different Sort of Time*, pp. 91–4.

a wide-ranging systems approach. The origins of the Hartwell study can be traced back to early 1950, when Mervin J. Kelly, president of the Bell Telephone Laboratories, advised the Chief of Naval Operations, Admiral Forrest P. Sherman, who was concerned about the vulnerability of U.S. ships to Soviet submarines, "to go to some place like MIT, since they have a lot of screwball scientists who will work on anything, and get a short study made...".[25] Zacharias accepted the directorship of the Hartwell summer study on the condition that the Navy should expect the scientists to come up not with a weapon but rather with some systematic thought about the broader subject of security of overseas transport. Zacharias saw himself as taking a systems engineering approach.[26]

Of the thirty-three scientists and engineers involved in Hartwell, ten, including Zacharias, were alumni of the MIT Radiation Laboratory or of the Los Alamos laboratory of the Manhattan Project. From these experiences, they brought an appreciation of the important role of interdisciplinary committees. Zacharias took the standards of the Radiation Laboratory as the standard against which all other research and development programmes should be judged. The repeated appearance of Radiation Laboratory and Los Alamos alumni on the committees advising the military in the 1950s has led some observers to speak of a Rad Lab and a Los Alamos mafia. Nine members had either Harvard or MIT backgrounds, but no one spoke of a Cambridge mafia.

Assigned the task of studying air defence, Project Charles, the new summer study, began meeting in February 1951 and submitted its final report in August 1951.[27] Not Zacharias but the well-liked and judicious F. Wheeler Loomis, head of the physics department at the University of Illinois and a former associate director of Radiation Laboratory, was chosen to head the project. Killian passed over Zacharias, a more obvious choice following his success with Project Hartwell, both because Zacharias did not get along well with Valley and because his views on the arrogance of Air Force officers were becoming well known.[28] Zacharias became associate director. (Years later, despite their differences, Zacharias commended Valley for having brought an unenthusiastic Air Force to the point where it would take ground-controlled air defence, not only air offence, seriously.)

Scrutiny of the membership of Project Charles and its consultants gives us a snapshot of the leadership — mostly East-Coast — of the industrial-university community cooperating with the military in the defence effort of the 1950s. The project had twenty-eight members, including eleven faculty or research associate members from MIT and five from other universities. The MIT contingent

---

[25] Goldstein, *A Different Sort of Time*, p. 99.

[26] Interview with Professor Jack Goldstein, Physics Department, Brandeis University, by Eric Rau, University of Pennsylvania, 26 November 1991.

[27] Project Charles, *Problems of Air Defense: Final Report of Project Charles* (Massachusetts Institute of Technology, 1951).

[28] Goldstein, *A Different Sort of Time*, p. 114.

included Gordon Brown, electrical engineer and guiding spirit of the Servo-mechanisms Laboratory there and a future dean of engineering at MIT; Forrester, then director of the Whirlwind computer project and soon to be head of the MIT Digital Computer Laboratory; R. Joyce Harman;[29] William R. Hawthorne, an MIT expert on aircraft propulsion; Albert Hill, physicist and head of MIT's Research Laboratory of Electronics; Malcolm Hubbard; J.C.R. Licklider, psychologist and later head of the Defense Advanced Research Project Agency's computer programme; H. Guyford Stever, aeronautical engineer and later Science Adviser to the President; William R. Weems; Valley; and Zacharias. Brown, Hawthorne, Licklider, Stever, Valley, and Zacharias were MIT professors; the others were research associates at laboratories there. Some members attended regularly; others were called on at intervals. Four participants were also members of ADSEC. Appointees from industrial and governmental laboratories included Ralph Johnson from Hughes Aircraft, Edwin Land, inventor and head of Polaroid Corporation, Harry Nyquist from Bell Telephone Laboratories, and Carl Overhage from Eastman Kodak Company. Marchetti represented the United States Air Force. Of the sixteen consultants in periodic attendance, six were from MIT. A liaison group of military officers from the United Kingdom, the United States, and Canada also attended sessions.

Consultants included Francis Bator, physicist from MIT; Lloyd Berkner, physicist from Associated Universities; Carl Kaysen, political economist from Harvard; C.C. Lauritsen, physicist from California Institute of Technology; John von Neumann, mathematician from the Institute for Advanced Study at Princeton; Louis Ridenour, physicist from the University of Illinois; Paul Samuelson, economist from MIT; James Tobin, economist from Yale; and Jerome Wiesner, electrical engineer from MIT and later Science Adviser to President Kennedy. Samuelson, Tobin, and Kaysen prepared reports on measures for industrial dispersion, a subject of special interest for Zacharias, who had headed a Project Charles committee on "passive defence".

The Project Charles final reports tell us about the procedures and deliberations of the group. Setting the stage, intelligence officers informed project members about an immediate threat from long-range Soviet TU-4 bombers and the likelihood that the Soviet Union "may be capable, by 1958, of producing supersonic high-altitude long-range bombers or guided missiles".[30] So alerted, the project group, drawing on its own engineering and scientific knowledge and that of its military, academic, and industrial consultants, recommended, on the organizational level, the integration of the air defence systems of the armed services, a rational objective sure to encounter massive opposition from the individual services. On the technical level, the members proposed a research and development agenda for short-range improvements in the existing air defence and a long-range agenda for achieving a new system. Realizing the immense

---

[29] I have not yet established the institutional affiliation of R. Joyce Harman.

[30] Project Charles, *Problems of Air Defense*, vol. 1, p. 3.

costs projected by the research and development agendas, the engineers and scientists in their final report declared that the extent and expense of the improved and new systems should be decided by the representatives of the people. "The problem of technical people", the report stated, "[is] to see that the funds are spent in ways that will be most effective".[31] This sentiment notwithstanding, congressmen and the administration, in the 1950s, tended to defer to scientists and engineers whom they considered experts.

The interdisciplinary study group delved into technical details and came up with an agenda stressing the importance of the development of automatic data-processing by using the digital computer. The report predicted that the electronic high-speed digital computer would have an important place as the coordinating apparatus in a centralized air defence; the report preferred a digital design over an analogue because the latter is circumscribed by its initial design, while a digital computer can be "quite thoroughly changed in function by the insertion of a new set of orders on a paper tape".[32] The report recommended a large digital computer that would embody and improve on the best features of the Whirlwind computer, which had already been field-tested by the Forrester team. Further, the committee called for an increase in the speed, capacity, and reliability of computer storage devices, both the electrostatic tubes then being used and the new three-dimensional array of ferromagnets, an invention for which Forrester and his staff became well known. Aware of new developments, the group recommended resorting to recently introduced transistors. Besides the information on automatic data processing, the report had lengthy sections on the need for new radar design and development. In accord with the concerns of Zacharias, the report conceded that air defence could not totally annihilate a large attacking force, so that passive defence measures, such as the dispersal of industry and urban air-raid shelters, were imperative because of the danger from atomic weapons.

Adhering to the recommendations of ADSEC and Project Charles, the Air Force, with limited participation by the US Army and Navy, funded the establishment of the Lincoln Laboratory as a mission-oriented laboratory focused on research and development for air defence. The SAGE air defence systems based on and designed around the digital computer, became the laboratory's principal project until 1958. In 1958 its first two computers became operational in a SAGE Direction Center at McGuire Air Force Base, New Jersey. These received aeroplane surveillance information from a radar network, calculated aircraft courses and speeds, drew on memory to locate interceptor aircraft and missiles, and processed information in order to transmit command and control information to interceptor aircraft and missiles. Because human operators were in the loop, the system was designated as semiautomatic. That same year, most of the engineers, scientists, and technicians in Lincoln Laboratory's Division 6, which

---

[31] Ibid., vol. 1, p. 112.

[32] Ibid., vol. 1, p. 112.

had designed and developed the computers in cooperation with International Business Machines (IBM), transferred to the newly founded MITRE Corporation, a Federal Contract Research Center. MITRE's mission was to develop improvements for the SAGE system and to perform the coordinating functions of a systems engineer. Eventually about twenty of the fifty Direction Centers that had been planned were built, but the deployment of intercontinental ballistic missiles substantially reduced the effectiveness of the SAGE air defence. Nevertheless, the research and development of SAGE was an enormously fruitful learning experience for a generation of engineers and scientists who contributed greatly to the development of the computer and to information processing.

# CONCLUSION

I have considered in some detail the contributions of external, interdisciplinary committees in the development of the Manhattan and the SAGE projects. Reflecting upon these, I propose several hypotheses:

1.  The interdisciplinary nature of the committees arose in part from the complexity of the projects. Only a broad array of disciplines and experiences could respond adequately to the multifaceted character of the projects.
2.  Academic engineers and scientists conceived of and became dominant members of the committees. The scientists and engineers from industrial laboratories who served on the committees had styles of research and development shaped by academic experiences and models.
3.  The committees designed research and development agendas and proposed organizational innovations; they did not simply review policies previously formulated.
4.  In the 1950s, decision and policy-making for research and development by interdisciplinary committees was far more common in academia than in industry, which tended towards hierarchical decision- and policy-making.

If the above hypotheses have validity, then I conclude that in the two major projects considered here, academics were more shapers than shaped; they infused military-funded projects with an academic style.

Finally, I should note that the committees considered in this essay were several among a family of external, interdisciplinary advisory committees that played major decision and policy-making roles in the field of research and development for the military-industrial-university complex from about 1940 to 1970. Three years after the formation of the Charles Summer Study, for instance, the Air Force Strategic Missiles Evaluation Committee, headed by John von Neumann, defined research agendas and instituted major organizational changes in the Air Force's Atlas Intercontinental Ballistic Missile project. Other committees, whose responsibilities transcended a single project, also had strong representation from the community of academic scientists. These included the Air Force's Scientific Advisory Board,[33] the Atomic Energy Commission's General Advisory Committee,[34] and the President's Science Advisory Committee.[35]

---

[33] Thomas A. Sturm, *The USAF Scientific Advisory Board. Its First Twenty Years*, (Washington, D.C.: 1967).

[34] Richard G. Hewlett and Oscar E. Anderson, Jr., *Atomic Shield, 1947–1952. A History of the United States Atomic Energy Commission* (Berkeley, CA: University of California Press, 1990), vol. 2.

[35] Gregg Herken, *Cardinal Choices. Presidential Science Advising from the Atomic Bomb to SDI* (New York: Oxford University Press, 1992).

# How Do We Know the Properties of Artefacts? Applying the Sociology of Knowledge to Technology

*Donald MacKenzie*

In this paper, I argue that the sociology of knowledge can help us understand how we come to know the properties of artefacts. By "properties", I mean their technical properties, rather than what we might call their "social meaning" (as in "the social meaning of the motor car"). Social meaning is, of course, fascinating, and it does affect our perception of technical properties, but it is obviously a sociological issue. Knowledge of technical properties, in contrast, might on the face of it seem a topic about which the sociologist could have nothing useful to say.[1]

The technical properties I am interested in are what we might call the "functional" or "behavioural" characteristics of artefacts: matters such as whether they work, how well they work, how safe they are. At the highest level of generality, these are vague matters. For specific technologies, however, they translate into more particular technical properties, such as the accuracy of a missile, the motor output of a turbine, the efficiency of a propeller, or the mean-time-between-failures of an electric component.

Our ordinary intuitions would tell us that these are indeed matters about which the sociologist could have nothing useful to say. Yet I believe those intuitions are

---

[1] For discussion of whether this is the case, see M. Mulkay, "Knowledge and utility: implications for the sociology of knowledge", *Social Studies of Science* (1973), **9**: 63–80; R. Kling, "Audiences, narratives, and human values in social studies of technology", *Science, Technology, and Human Values* (1992), **17**: 349–65; K. Grint and S. Woolgar, "Computers, guns, and roses: what's social about being shot?", ibid., pp. 366–80; and R. Kling, "When gunfire shatters bone: reducing sociotechnical systems to social relationships", ibid., pp. 381–5.

247

mistaken. In arguing this, I shall draw on the sociology of scientific knowledge, because that is the field where the sociological study of technical matters is most advanced. I should emphasize, however, that drawing on the sociology of scientific knowledge in this way does not imply that "science" and "technology" are the same (they are not), nor that technology is applied science (it is not).[2]

In making the argument, I use some empirical examples. I am not claiming these examples to be in any sense typical; rather, they are chosen for the simple and (I hope) vivid way in which they illustrate the argument.[3] Nor, on the other hand, are they unique. In fact, at the end of the paper I speculate that processes akin to those discussed here may actually be pervasive in situations of technological change.

## THE SOCIOLOGY OF KNOWLEDGE

Two preliminary remarks about the sociology of knowledge are necessary. First, for the sociologist of knowledge, "knowledge" means any shared belief system, not necessarily correct belief. Indeed, the modern sociology of scientific knowledge argues that sociological explanation should not be influenced by our assessment of the validity of the beliefs in question. This symmetry principle is the core of what David Bloor has called the "strong programme" of the sociology of knowledge, in contrast to a "weak" programme that would restrict sociological explanation to beliefs regarded as false, irrational, or in other ways inadequate.[4]

In studying knowledge of the properties of artefacts, the "strong programme" sociologist of knowledge therefore has to reject the presupposition that:

> A rational, objective knowledge is there if it is sought ... though groups may ... choose not to seek it, or may ignore it when it is presented to them. But this does not make the knowledge "social" — it makes the project controllers careless, wicked, lacking in

---

[2] The relevance of the sociology of knowledge for the study of technology has been argued most forcibly in Trevor J. Pinch and Wiebe E. Bijker, "The social construction of facts and artefacts: or how the sociology of science and the sociology of technology might benefit each other", *Social Studies of Science* (1984), **14**: 339–441. On the relationship between science and technology, see, for example, Barry Barnes and David Edge (eds.), *Science in Context. Readings in the Sociology of Science* (Milton Keynes: Open University Press, 1982), part three.

[3] The empirical work drawn on here was supported by the Nuffield Foundation (grant SOC/442), the Economic and Social Research Council (grants R000290008, WA35250006, Y307253006 and R000234031), the Science and Engineering Research Council (grant GR/J58619), and by the Joint Committee of the last two bodies (grant GR/H74452).

[4] The "strong programme" was first spelt out in D. Bloor, "Wittgenstein and Mannheim on the sociology of mathematics", *Studies in the History and Philosophy of Science* (1973), **4**: 173–91. It has been intensely controversial: see Bloor's "Afterword: attacks on the strong programme", to the second edition of his *Knowledge and Social Imagery* (Chicago: Chicago University Press, 1991), pp. 163–95.

integrity and a disgrace to their profession. It does not make ... the facts subject to social matters.[5]

The problem with this asymmetric formulation can be seen in sharpest focus in the study of scientific controversy. Historians and sociologists have found cases, even in heartland sciences like physics, where what one group of (careful and honest) people takes as patently rational objective knowledge is dismissed by others (equally careful and honest) as self-evident error and folly.[6] Sociologists of scientific knowledge have seen their task as being to explain how different groups of people come to put forward different knowledge claims, rather than to adjudicate between these claims, or to put forward an analysis premised upon the superiority of one set of claims. Faced with technological controversy, the task of the sociologist of technical knowledge is surely the same.

The second preliminary remark about the sociology of knowledge is that the "social" is not the same as "context", which is how the wider society or wider culture is sometimes referred to in the literature on scientific or technological change. When I use terms like "social convention" or "social interest", the "societies" being referred to will frequently be communities of technologists, not larger groupings such as nations. A sociological analysis is not the same as what in the old debates about the history of science used to be called an "externalist" perspective. The relative roles of social processes inside and outside scientific or technological communities is a contingent, empirical matter. Take, for example, the paper presented by Thomas Kuhn to the 1961 Oxford Conference on Scientific Change. Kuhn's was not an externalist analysis of science (the wider society plays relatively little role in it), but it was still deeply sociological in its focus on the relations between scientific communities and their knowledge.[7]

## HOW DO WE KNOW THE PROPERTIES OF ARTEFACTS?

When we consider how people come to know the properties of artefacts, three processes suggest themselves:

— we know them by *authority*: people whom we trust tell us what these properties are;

---

[5] The quotation is from M.C. Duffy's review of D. MacKenzie, *Inventing Accuracy. A Historical Sociology of Nuclear Missile Guidance* (Cambridge, Mass.: MIT Press, 1990) in *Archives internationales d'histoire des sciences* (1992), **42**: 346–49, on p. 348.

[6] See, for example, Steven Shapin, "History of science and its sociological reconstructions", *History of Science* (1982), **20**: 157–211, and Harry Collins and Trevor Pinch, *The Golem. What Everyone Should Know About Science* (Cambridge: Cambridge University Press, 1993).

[7] T.S. Kuhn, "The function of dogma in scientific research", in A.C. Crombie (ed.), *Scientific Change* (London: Heinemann Educational Books, 1963), pp. 347–69.

— we know them by *induction*: we learn the properties of artefacts by testing or using them;
— we know them by *deduction*: we infer their properties from theories or models (for example, but by no means exclusively, by deduction from scientific theories, such as Newtonian physics or Maxwellian electromagnetism).

I am not claiming that authority, induction, and deduction are independent of one another: indeed, my argument is precisely that they are not, and that, in particular, matters of authority shape the processes of induction and deduction. Nor is this list necessarily exhaustive: all I suggest is that any plausible additions to it also involve social processes.[8]

Let me begin with authority, about which I shall be briefest because it patently is a sociological matter. Where we are located in society — in which social class, in which gender, in which ethnic group, and so on — shapes whom we trust. Furthermore, trust is intrinsically multi-dimensional: our assessment of the knowledge claims of others is affected by the entirety of our relations to them.

The heterogeneity of that entirety of relations is, for example, the topic of some of the most interesting feminist research on science, technology, and medicine. Thus Coral Lansbury starts with an observation that several historians have made: the striking prominence of women in the nineteenth-century British movement against vivisection. It is easy to assume that this prominence came about because women were particularly prone to a sentimental attachment to animals. Lansbury, however, argues differently, claiming that:

> ... it was not the plight of animals alone which stirred them [women] to anger, but their own. The issues of women's rights and antivivisectionism fused at a level which was beyond conscious awareness, and animals were increasingly seen as surrogates for women who read their own misery into the vivisector's victims.[9]

Of course, unconscious motivation is, by definition, hard to demonstrate; yet it is plausible that women's experience of degrading treatment by male doctors,

---

[8] For example, Svante Beckman (personal communication) suggests inference from *design* as an addition. This is certainly plausible: if we know that an artefact has been designed to be a mousetrap, we may well be confident (without having to try it out) that it catches mice. Inference from design could be regarded as a species of deductive inference, and so at least some of the following remarks about deduction apply. Furthermore, if the "designer" and "user" are different people, then issues of trust and authority are clearly involved in the inference.

[9] Coral Lansbury, "Gynaecology, pornography and the antivivisection movement", *Victorian Studies* (1985), **28**: 413–37, quotation on p. 426; thanks are due to John Henry for pointing me to this article. See also Richard D. French, *Antivivisection and Medical Science in Victorian Society* (Princeton, NJ: Princeton University Press, 1975), and Mary Ann Elston, "Women and anti-vivisection in Victorian England, 1870–1900", in Nicolaas A. Rupke (ed.), *Vivisection in Historical Perspective* (London: Routledge, 1990), pp. 259–94.

especially by gynaecologists (experience gained as patients and also in some cases as pioneering women doctors) should have led them to distrust vivisecting medical men. Lansbury suggests that sexual images, in particular pornographic images of women bound and forced into submission, haunted the debate over vivisection. There was, for example, suspicion that vivisectors gained sexual pleasure from their treatment of animals, and even widespread popular belief that Jack the Ripper was a vivisecting London University surgeon "who had extended his research from animals to women".[10]

The breakdown of trust exemplified by these suspicions had cognitive effects. Anti-vivisectionists typically rejected the validity of the knowledge gained as a result of vivisectionist research (paradoxically, this rejection sometimes took the form of an explicit rejection of the inference from what was true of animals to what was true of humans). They were predisposed instead to believe in the effectiveness of forms of medicine, such as homeopathy, that did not rely on this form of induction.

I lack the historical knowledge to assess the empirical validity of Lansbury's argument, and in any case it deals with medicine rather than technology. But, in general, there is good reason to believe that if, for *whatever* reason, trust does not exist or breaks down, then cognitive authority vanishes too. That, for example, is the message of the best of the literature on risk perception and of that on the public reception of science. How safe we take a technology to be is affected by our trust, or lack of it, in those who run it; whether we believe "experts" depends on our relations to them.[11]

# INDUCTION FROM TESTING

Let me move on from "authority" to "induction", first of all in the form of induction from the testing of technologies. Induction is reasoning from particular cases to a general conclusion, or from past cases to future cases. That is precisely what we do when making inferences on the basis of testing.

What is social about induction? It is perfectly plausible to argue that human beings come into the world equipped with natural inductive propensities,[12] and in that sense it can be argued that induction is a psychological rather than sociological matter. One can certainly imagine that a feral child, brought up in complete isolation from other human beings, might reason inductively about the properties of such tools as he or she came up with. However, in all actual human

---

[10] Lansbury, "Gynaecology, pornography and the antivivisection movement", p. 431.

[11] See, for example, Mary Douglas and Aaron Wildavsky, *Risk and Culture. An Essay on the Selection of Technical and Environmental Dangers* (Berkeley, CA.: University of California Press, 1982), and Brian Wynne, "Misunderstood misunderstanding: social identities and public uptake of science", *Public Understanding of Science* (1992), **1**: 281–304.

[12] Barry Barnes, "Natural rationality: a neglected concept in the social sciences", *Philosophy of the Social Sciences* (1976), **6**: 115–26.

societies our natural inductive propensities are channelled by the culture into which we are born. Thomas Kuhn has provided us with a simple but striking analysis of this process.[13] He points out that inductive learning rests on what he calls "similarity relationships". No two perceptible objects or events are ever identical. In order to gain generalizable empirical knowledge, or to reason from past events to future ones, we have to learn which similarities are important, and which differences are unimportant.

Kuhn's example is of a child walking with its parent. The parent points at a bird and says: "Johnny, that's a swan." Johnny points at another bird and says "swan". "No", says the parent, "it's a goose". Gradually, in that way, the child learns to differentiate between swans, geese, and ducks — and learns to do so reliably, even though no two swans are identical, and no parent could offer an exhaustive verbal definition of "swan". These learnt similarity relationships then form the basis for inductive generalizations, such as "geese hiss, ducks don't".

The example is simple, but here we are at the core of Kuhn's philosophy of science:

> Need I now say that the swans, geese, and ducks which Johnny encountered during his walk with father were what I have been calling exemplars? Presented to Johnny with their labels attached, they were solutions to a problem that the members of his prospective community had already resolved. Assimilating them is part of the socialization process by which Johnny is made part of that community and, in the process, learns about the world which the community inhabits.[14]

Similarity relationships are imposed by us upon nature, rather than (at least in any simple sense) by nature on us. In a wholly non-pejorative sense, similarity relationships are social *conventions*.[15] Different human communities, for example, have different classifications of birds. Each classification "works" from the point of view of the purposes of the community that adheres to it. There is, as Kuhn famously argued, no entirely context-independent way of demonstrating the superiority of one classification, or one "paradigm", over others.

Kuhn's ideas have been applied to technology in a variety of ways, some quite complex and ambitious.[16] To grasp what is social about induction from the

---

[13] Thomas S. Kuhn, "Second thoughts on paradigms", in Kuhn, *The Essential Tension. Selected Studies in Scientific Tradition and Change* (Chicago: University of Chicago Press, 1977), pp. 293–319.

[14] Kuhn, "Second thoughts on paradigms", p. 313.

[15] See Barry Barnes, *T.S. Kuhn and Social Science* (London: Macmillan, 1982), especially chapter 2.

[16] See, for example, Edward W. Constant II, *The Origins of the Turbojet Revolution* (Baltimore, Md.: Johns Hopkins University Press, 1980); Giovanni Dosi, "Technological paradigms and technological trajectories: a suggested interpretation of the determinants of technical change", *Research Policy* (1982), **11**: 147–62; and Rachel Laudan (ed.), *The Nature of Technological Knowledge. Are Models of Scientific Change Relevant?* (Dordrecht: Reidel, 1984).

results of testing technologies, however, we need reflect only on one particular, restricted type of similarity relationship: that between testing and use. Typically, when we test technologies, our goal is to infer how those technologies will behave in use.[17] So the similarities and differences between testing and use become crucial. Just as no two swans are wholly identical, it is always possible to find ways in which tests differ from use. Inductive inference from the former to the latter therefore rests upon the *judgement* that testing is "sufficiently like" use to allow inferences to flow, or at least that the differences are "sufficiently well understood" as not to form a barrier to inference.

That there is indeed a judgement involved here can be seen most clearly in cases where the similarity between testing and use is contested. Consider, for example, the testing of nuclear inter-continental ballistic missiles. The nations that possess such missiles test them by firing a sample of missiles, without their nuclear warheads, on a test range, for example the US range from Vandenberg Air Force Base in California to Kwajalein Atoll in the Marshall Islands. From that testing, an accuracy figure for the missile — a "circular error probable" — is inferred. This in turn is taken as providing knowledge of what will happen in wartime flights: in attacks on, say, Moscow or Baghdad.[18]

In the United States in the early 1980s, the validity of this inference from testing to use was sharply disputed by critics who challenged the similarity relationships underpinning it. For example, critics argued:

— that flying north over the Arctic to Russia was different from flying west over the Pacific;
— that repeated firing on the same test range was different from one-off wartime firing;
— that test missiles were different from missiles deployed in silos.

On the other hand, those who defended missile accuracies as genuine facts countered that these differences were either trivial or well understood. Testing was sufficiently like use, they claimed.

No clear-cut resolution of this debate was reached: opposed judgements remained just that. Nor, indeed, was there ever a definitive resolution of a previous debate, in the early 1960s, which focussed on whether intercontinental ballistic missiles would work at all, in the basic sense of whether their nuclear warheads would actually explode. Critics had pointed out that intercontinental

---

[17] There are forms of testing which do not have this goal, at least not directly: for example, some tests are informed by theoretical models of the technology in question, and their goal might be to find the value of a particular parameter in the model. For the sake of simplicity, I restrict the discussion here to tests where the intention is direct inference to use.

[18] References to the original debates discussed here can be found in D. MacKenzie, "From Kwajalein to Armageddon? Testing and the social construction of missile accuracy", in David Gooding, Trevor Pinch, and Simon Schaffer (eds.), *The Uses of Experiment. Studies in the Natural Sciences* (Cambridge: Cambridge University Press, 1989), pp. 409–35; also in MacKenzie, *Inventing Accuracy*, chapter 7.

ballistic missile tests were conducted without warheads on board, and that normally the warheads in test explosions were stationary. They argued that testing missiles and warheads separately, in this way, was not a reliable basis for inference as to what would happen when, in use, missile and warhead were joined together.

The radically sceptical point of view that we have no assurance that nuclear intercontinental ballistic missiles will actually work was advanced by leading figures in the United States defence establishment in the early 1960s, such as the then Chief of Staff of the US Air Force, General Curtis LeMay. It has occasionally resurfaced since, as when a correspondent asked in the *Bulletin of the Atomic Scientists* in 1986: "Does a test conducted dead still at the bottom of a hole provide reliable assurance that the same weapon will work after travelling several thousand miles per hour in a re-entry vehicle through extremes of temperature?"[19] The correspondent felt the answer was "No", but the judgement of the community of nuclear weapons designers in the United States now appears to be that these differences between testing and use are not crucial.

The history of these two controversies reveals a clear social patterning to the differences in judgement involved. For example, proponents of the nuclear bomber were prominent in both forms of criticism of inferences from missile testing to use. The different positions in the controversy served different social interests, and when the relevant circumstances changed, the course of the controversy altered accordingly.[20]

Of course, the testing of nuclear intercontinental ballistic missiles is a quite untypical case, because, fortunately, they have never been used, and there are major legal and political constraints on how they can be tested.[21] However, I gain at least a little confidence about the generalizability of the analysis given here by reading the two historians of technology who have focussed most centrally on testing: Edward Constant and Walter Vincenti. (It is worth noting, in passing, that there appears to be a curious neglect of this topic. Scientific experiment has commanded much attention, but relatively little has been paid to the testing of technology, which is in a sense its analogue.)

---

[19] A.L. Meyers, "Nuclear testing", *Bulletin of the Atomic Scientists* (August–September 1986), **43**: 66–67.

[20] Thus scepticism about missile accuracy figures was welcomed in the period 1980–2 by critics of the new US intercontinental ballistic missile, MX, because it undermined the argument that existing Minuteman missiles were vulnerable to attack from highly accurate Soviet missiles. When, in 1983, the Reagan Administration had to accept deployment of MX in existing Minuteman silos, this audience for the sceptical argument largely disappeared, and the controversy itself passed from public view.

[21] Thus the Partial Test-Ban Treaty of 1963 prohibits nuclear testing in the atmosphere, and thus rules out what some critics have seen as the only "realistic" way of testing nuclear missiles: i.e. with live warheads.

Vincenti's work on the testing of air propellers in the early twentieth century shows the centrality of similarity relationships: "laws of similitude", he calls them. Aeronautical engineers tested scale models, not full-scale prototypes: they had to learn the similarity relationships that would permit reliable induction from test to use. They had to learn the phenomena that mattered and also (as in all inductive inference) the phenomena that did not matter: for example, they learned that they could neglect the viscosity and compressibility of the air, the elastic bending of the propeller, and the difference between the enclosed space of a wind-tunnel and the open space of the skies.[22]

As Constant points out in his discussion of the Prony brake (a means of measuring motor output, used, for example, to test turbines), this kind of process is a communal as well as an individual one. Communities of technologists typically come to agree on what similarities matter and what differences do not in technological testing. Indeed, to the extent that they agree on what constitutes valid testing, they simultaneously create themselves as communities: communal authority, and not just individual trust, now underpins knowledge claims.[23]

If, however, those involved do not agree on the validity of different test procedures, we may have two or more viewpoints on what kind of testing is most like use, often linked to different perceptions of which kind of artefact is best. This seems to have been the case, at least in part, in the controversy in the early 1960s between proponents of the US Army Ordnance Corps's rifles and sup-porters of the Armalite Company's AR-15 rifle.[24] In such circumstances, each side can appeal to test results, can consider its knowledge rational and objective, and may even dismiss the other as careless or wicked.

Even when competing forms of testing do not exist, a sociological analysis is still possible. Different interests may be served, as they were in the missile case, by criticism of testing and by its defence. There may, furthermore, be a typical pattern to the confidence shown in the results of technological testing. That pattern I would call the "certainty trough". On the vertical axis in my figure lies uncertainty; on the horizontal axis, "social distance" from the production of knowledge.

The figure shows three typical combinations of uncertainty and social distance. On the left are those directly involved in testing. These tend to be loyal, insider members of their technological community, but their intimate

---

[22] Walter Vincenti, "The air-propeller tests of W.F. Durand and E.P. Lesley: a case study in technological methodology", *Technology and Culture* (1979), **20**: 712–51.

[23] Edward W. Constant, "Scientific theory and technological testability: science, dynamometers, and water turbines in the 19th Century", *Technology and Culture* (1983), **24**: 183–98.

[24] See, for example, James Fallows, "The American Army and the M-16 rifle", in D. MacKenzie and J. Wajcman (eds.), *The Social Shaping of Technology* (Milton Keynes: Open University Press, 1985), pp. 239–51.

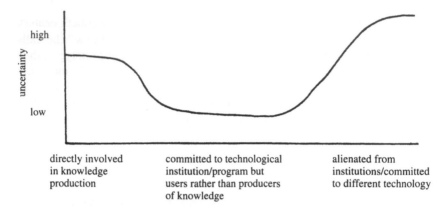

directly involved          committed to technological          alienated from
in knowledge               institution/program but             institutions/committed
production                 users rather than producers         to different technology
                           of knowledge

**Figure 2**   The Certainty Trough

experience of the vagaries of knowledge production gives them access to sources of uncer-tainty barred to others. I found several examples of this when studying missile accuracies: insiders, not public critics, would tell me of things like their worries about the significance of climate. Vandenberg Air Force Base enjoys a pleasantly temperate Californian coastal climate, but it was very different to have to align a missile guidance system — in old systems, a delicate manual operation — in the mid-West winter. Vincenti's Everett Lesley, to take another example, would no doubt have told us about the importance of precisely how one should varnish one's test propellers and about what could go wrong otherwise.[25] This kind of "insider uncertainty" is the analogue, for technology, of the uncertainty manifest among those intimately involved in frontier science, especially controversial science, who have come to understand the fallible and skill-dependent nature of scientific knowledge.[26]

On the right of the figure are those (when they exist: they do not seem to in many cases, such as those of air propeller or turbine testing) who are alienated from the institutions responsible for the technologies in question: in the missile case, they include both those hostile to the nuclear arms race as a whole, and at least some of the more single-minded proponents of the bomber. In their case, precisely as the literature on risk perception would suggest, a lack of trust in institutions has bred radical scepticism about the knowledge these institutions produce.[27]

In the middle of the figure are to be found the inhabitants of the "certainty trough". These are those loyal to institutions but without direct involvement in the process of knowledge production, who therefore tend to lack access to the ways in which testing and use can be seen as different. They include senior

---

[25] Vincenti, "Air-propeller tests".

[26] See, for example, Collins and Pinch, *The Golem*.

[27] Douglas and Wildavsky, *Risk and Culture*.

generals, high-level political decision-makers, top managers, and the like. I use the phrase "certainty trough," not "uncertainty trough", because in many forms of technological activity (not least those having to do with nuclear missiles), it is their certainty, not their uncertainty, that is the worrying thing.


# INDUCTION FROM USE

Nuclear missiles are, to repeat, unusual in being tested but (so far) not used. Normally, we use technologies as well as test them, and induction from use is certainly a major source of knowledge of the properties of artefacts. Surely that is unproblematic: surely use straightforwardly tells us, for example, whether or not a technology works?

Not always. Consider the Gulf War of 1991, when many military systems, long restricted to testing alone, were used for the first time in war. That did seem to bring unproblematic knowledge. After a decade of the "Star Wars" debate about whether defence against ballistic missiles was feasible, suddenly there, on our television screens, was the American Patriot missile system successfully defending Israel and Saudi Arabia against Saddam Hussein's Scuds. Whatever doubts might have been entertained about the testing of Patriot, surely here was unequivocal evidence that it worked? "42 Scuds engaged, 41 intercepted", said President Bush.[28] Subsequently, however, the question of Patriot's Gulf War anti-ballistic missile efficacy became embroiled in controversy. Whether Patriot worked — indeed, what it was for Patriot to have worked — was fiercely contested. Several forms of evidence are involved in this debate, but let me focus on the visual evidence: what eyewitnesses or television viewers saw.

There have been at least three interpretations of this visual evidence, and the last two (at least) are still in contention. The first interpretation is that typical of the television reporting at the time. Satisfying bangs and explosions in the sky were heard and seen as successful interceptions. Residual explosions at ground level were typically accounted for as either spent Patriots exploding or Scuds successfully being diverted away from their targets to wasteland. Patriot was seen as almost always successful.

The second interpretation is that of Theodore A. Postol, Professor of Science, Technology and National Security Policy at the Massachusetts Institute of Technology, and other critics of Patriot.[29] These critics argue that if one examines the video recordings of supposed interceptions carefully, frame by

---

[28] Quoted in W. Safire, "The great Scud-Patriot mystery", *New York Times*, 7 April 1991, p. A25.

[29] See, for example, Theodore A. Postol, "Lessons of the Gulf War experience with Patriot", *International Security* (Winter 1991–2), **16**: 119–71; Postol, "Correspondence. Patriot experience in the Gulf War", ibid. (Summer 1992), **17**: 225–40.

frame, one sees Patriot almost always missing its target, or perhaps intercepting the large Scud fuel tank and missing the small but important warhead. A Scud falling in wasteland or failing to explode is, for the critics, no evidence of successful interception: the Scud is an inherently inaccurate and unreliable missile. According to critics, Patriot's success was at best poor: indeed, perhaps there were no successful intercepts.

The third interpretation of the visual evidence is that of the defence against this criticism.[30] This defence has, for example, queried the visual adequacy of broadcast videotape. Defenders of Patriot have argued that broadcast videotape's low frame rate, combined with the very high relative speed of Scud and Patriot, can make what is really a hit seem like a miss by not capturing the precise instant of interception.[31] Defenders of Patriot claim that one would need the systematic use of specialized ultra-fast test-range cameras, rather than the *ad hoc* use of ordinary video cameras, to know visually whether Patriot was working or not.

Again, this is an unresolved debate; it is also, as I noted above, one in which both critics and defenders draw upon many sources of evidence other than the visual record. Again, too, different potential resolutions of the debate would have favoured or disadvantaged particular interests. The perception of Patriot as an effective anti-ballistic missile defence seems to have enhanced sales of the system (Saudi Arabia and other nations having subsequently made major purchases), and its apparent success was taken as a powerful argument in favour of the Strategic Defense Initiative. Conversely, critics of Star Wars could be said to have had an interest in Patriot's being seen as a failure, as could Democratic politicians wishing to criticize the record of President Bush's Republican Administration.[32]

## DEDUCTION

For all the importance of induction from testing and use, our knowledge of the properties of artefacts is seldom purely inductive. We also infer the properties of artefacts from theories and models. When we are dealing with advanced tech-

---

[30] Defences include R.A. Drolet, "PEO air defense response to Patriot criticisms", *Inside the Army*, 9 December 1991, pp. 5–6; C. Zraket, "Patriot gave stellar Gulf performance", *Defense News*, 9 December 1991, pp. 31–2; and R.M. Stein, "Correspondence. Patriot experience in the Gulf War", *International Security* (Summer 1992), **17**: 199–225.

[31] Postol, in response, admits that low frame rate can be a problem, but argues that this effect is not sufficiently large to account for the miss distances suggested by the videotapes.

[32] As with the controversy over missile accuracy, changing political circumstances (notably the Democratic victory in the 1992 Presidential Election and the subsequent abandonment or postponing of the most ambitious forms of the Strategic Defense Initiative) seem to have removed some of the audience for the debate.

nologies in industrialized societies, these theories and models are often either scientific ones or specialized engineering ones. But they are by no means always so: the properties of artefacts can also be inferred from, for example, magical or religious beliefs.

One basic point about all such deductive knowledge is that, like inductive knowledge, it too rests upon similarity judgements. For deduction to be valid, the situation in question must be sufficiently like those situations to which the theory, model, or belief system is known to apply, or to which it has successfully been applied in the past. Whether the situations are sufficiently alike is, however, often problematic. To claim that they are not is one of the main ways in which scientific expertise can be challenged in technological controversies: the particular way in which scientific knowledge has been applied in any particular case can be contested without questioning the authority of science *per se*.[33] And the issue is pertinent not only for those embroiled in such controversies. Engineers have frequently believed that the technological situations they have to deal with are too unlike those for which apparently relevant scientific theories have been developed. This is a major reason why they have developed more specialized, situation-specific engineering theories.[34]

Here, however, I want to raise a different issue about deductive inference: the question of its internal structure, rather than that of its applicability. This internal structure is the core of the special status enjoyed by deductive reasoning, because it is seen as absolutely compelling, rather than merely probabilistic, like inductive inference. The steps of deductive inference are seen as incontestable, as logically or mathematically self-evident. If its premises are granted, the conclusion of a properly formed deductive argument *must* follow.

The special status of deductive inference has meant a sharply growing role for it recently as a source of knowledge about artefacts, particularly of knowledge about computer systems. Traditionally, knowledge about computer programs and computer hardware has been gained inductively, by testing them and by experience gained in using them. However, particularly where computers are crucial to safety and security, it has been argued influentially that testing is not good enough: that computer systems are too complex ever to be tested exhaustively. As one leading computer scientist, Edsger Dijkstra, put it in 1972, in words that have often been repeated since: "program testing can be a very effective way of showing the presence of bugs, but it is hopelessly inadequate for showing their absence". What "mathematicizers" like Dijkstra wanted — and what some of them have subsequently claimed to have provided — is deductive proof that programs or

---

[33] See, for example, Barnes and Edge, *Science in Context*, part five.

[34] For the significance of these engineering theories, see, for example, Edwin T. Layton, Jr., "Mirror-image twins: the communities of science and technology in 19th century America", *Technology and Culture* (1971), **12**: 562–80.

computer hardware designs are correct implementations of their specifications: "the only effective way to raise the confidence level of a program significantly is to give a convincing proof of its correctness".[35]

This enterprise is fascinating. Despite the aura of certainty and absoluteness surrounding logical inference and mathematical proof, for the sociologist of knowledge deductive argument is simply another challenge, albeit a harder one than inductive argument. Indeed, key to the original formulation of the strong programme of the sociology of knowledge was David Bloor's argument that logic and mathematics are not exempt from sociological explanation. Drawing on the history of mathematics and, especially, the history of logic, Bloor has shown that there are major variations in what kinds of deductive inference are taken as valid: they are variations that in some cases at least may be socially patterned.[36]

In 1987, colleagues and I drew upon this sociological work in making a prediction about the use of deductive inference to prove the properties of computer systems. We argued that despite the apparently absolute nature of mathematical proof, this move of proof into what would inevitably become a commercial and regulatory arena would lead to a court of law having to rule on what a mathematical proof is.[37] That prediction was nearly borne out in 1991, when litigation broke out in Britain over a microprocessor chip called VIPER (Verifiable Integrated Processor for Enhanced Reliability). VIPER was the first commercially available microchip with a claimed mathematical proof of the correctness of its design. At issue in the litigation was precisely whether that claim of proof was or was not correct. Following dispute over this point within the computer science community, the Ministry of Defence, which had made the claim, was sued under the Misrepresentation Act by one of VIPER's licensees, a company called Charter Technologies, Ltd. The case failed to come to court only because Charter went bankrupt before the High Court hearing.

The details of the VIPER case have been reported elsewhere.[38] However, a couple of speculations based on it and other more recent research may indicate how a "sociology of proof" in this area might be developed. The first arises from the way in which the question of proof is, in practice, often intimately bound up

[35] E.W. Dijkstra, "The humble programmer", *Communication of the Association for Computing Machinery* (1972) **10**: 859–66, quotations on p. 864. This kind of statement presaged an interesting clash of technical communities, with proponents of program testing rallying to the defence of inductive knowledge. Unfortunately, space prohibits exploration of this clash.

[36] See especially, D. Bloor, *Wittgenstein. A Social Theory of Knowledge* (London: Macmillan, 1983).

[37] E. Peláez, J. Fleck, and D. MacKenzie, "Social research on software", paper presented to Economic and Social Research Council, Programme on Information and Communi-cation Technologies, Manchester, December 1987, p. 5.

[38] D. MacKenzie, "The fangs of the VIPER", *Nature* (8 August 1991), **352**: 467–8.

with that of trust. Many proofs of computer system correctness, and indeed some proofs in mathematics, are simply so long that they are difficult or impossible fully to comprehend or even to read in full. For example, at the time of the litigation, the main VIPER proof consisted of 7 million computer-executed primitive deductive inferences. Andrew Wiles's celebrated recent claimed proof of Fermat's last theorem, though human- not computer-generated, is several hundred pages long. In mathematical group theory, the proof of the completeness of the classification of the finite simple groups is a collective effort stretching over upwards of 10,000 pages.

Among the issues of trust (and there are several[39]) that are crucial to the reception of such proofs is whether to trust human beings, who may be full of insight but may also be liable to wishful thinking, or to trust computers, which are typically seen as untiring but blind. Attitudes to this issue of trust may be influenced by wider attitudes to the relationship between humans and machines. This may (I speculate) underpin differences of opinion among mathematicians on the acceptability of computer-assisted proofs, such as that of the four-colour theorem in graph theory. For example, dismissing these as mere "quasi-proofs", one leading mathematician wrote: "we cannot possibly achieve what I regard as the essential element of a proof — our own personal understanding — if part of the argument is hidden away in a box".[40]

A second issue is that there may well be differences in the criteria that different disciplinary communities — such as logicians, mathematicians, physicists, and engineers — use in judging whether a particular mathematical argument is indeed a proof. For the logician, proof is "formal proof": a finite sequence of formulae in which each formula is either an axiom or derived from previous formulae by the "mechanical" application of sound logical laws. The mathematician, if pressed, might also offer such a definition, but in practice few mathematical proofs take this form, and the criterion of "understanding" — the "surveyability" of claimed proofs — seems to be of considerable importance.[41] The physicist may regard formality (in the logician's sense) as unnecessary or unachievable, but may require that

---

[39] Issues of trust are not restricted simply to long proofs. Thus one mathematician, interviewed for this work by my colleague Dr Anthony Dale, reported that many of his own mathematical papers were accepted "suspiciously quickly"! He believed that "some referees looked only at the author and the theorem" before accepting the paper: that is to say, they placed their trust in the reputation of the author and the plausibility of the theorem, without examining the details of the putative proof.

[40] F.F. Bonsall, "A down-to-earth view of mathematics", *American Mathematical Monthly* (1982), **89**: 8–15, on p. 13. For broader discussion of the issue, see John Forgan, "The death of proof", *Scientific American* (October 1993), **269**: 74–82.

[41] One interest of computer-generated proofs is that they may force choice between the criterion of formality and that of surveyability: thus they are nearly always formal, but often not surveyable, at least not easily so.

deductive reasoning always preserve "the physical interpretability of the equations throughout the derivation".[42] For the engineer, deduction from axioms might not be regarded as important, and rigorous checking of all possible cases might be seen as constituting mathematical proof.[43]

One of the interesting aspects of computer science, as a discipline, is that recruits to it come from all four of these disciplinary backgrounds (as well as from others). Although the logicians' definition of proof has undoubtedly been dominant within computer science, other perspectives exist and have been drawn upon, for example in the VIPER controversy or in the new UK Ministry of Defence standard governing safety-critical software in defence equipment.[44] A current research project is investigating such differences in criteria of proof and whether they do indeed have disciplinary or other social bases.[45]

To repeat, the above suggestions for a "sociology of proof" in this area are speculations rather than findings. If, though, they turn out to have any foundation, they do point to the conclusion that, despite the aura of absolutism surrounding deductive proof, our deductive knowledge of artefacts may be no more immune to sociological explanation than our inductive knowledge is.

## CONCLUSION

As I said at the start of this paper, the examples I have discussed have no claim to being typical. Yet it has been my experience that whenever I have studied technological change, I have found at least a degree of controversy over our knowledge of the properties of artefacts. Often it has not been public

---

[42] Eric Livingston, *The Ethnomethodological Foundations of Mathematics* (London: Routledge and Kegan Paul, 1986), p. 188; see also Arthur Jaffe and Frank Quinn, "'Theoretical mathematics': toward a cultural synthesis of mathematics and theoretical physics", *Bulletin of the American Mathematical Society* (1993), new ser. **29**: 1–13.

[43] In the VIPER controversy, for example, one issue was whether a method of exhaustive checking called "intelligent exhaustion" could be regarded as proof. For the method, see C.H. Pygott, *Formal Proof of Correspondence Between the Specification of a Hardware Module and its Gate Level Implementation* (Malvern, Worcs.: Royal Signals and Radar Establishment, November 1985), report no. 85012.

[44] The chief view of proof alternative to that of the logician has been, interestingly, a "sociologized" version of the mathematicians' criterion of surveyability: R.A. DeMillo, R.J. Lipton, and A.J. Perlis, "Social processes and proofs of theorems and programs", *Communications of the Association for Computing Machinery* (1979), **22**: 271–80. For a discussion, see D. MacKenzie, "Negotiating arithmetic, constructing proof: the sociology of mathematics and information technology", *Social Studies of Science* (1993), **23**: 37–65.

[45] "Studies in the sociology of proof", jointly with Alan Bundy and Anthony Dale (Economic and Social Research Council grant R000234031).

controversy; sometimes it does not even find its way into the technical literature. Interviewing — oral history — may be the only way of uncovering it.[46] But it may be that, in situations of technological change, controversy over the properties of artefacts is pervasive. Technological change, after all, often disturbs established social institutions and patterns of social interests. If our knowledge of the properties of artefacts is indeed socially contingent, as this paper has argued, then it would not be surprising if the conflicts generated by technological change typically involve disagreement over technical knowledge.

I would also argue, however, that knowledge of the properties of artefacts is no less social when there are no such controversies to examine. It is true that stable technologies are different from situations of technological change. Associated with stable technologies, we can usually find stable knowledge of the properties of artefacts, agreed authorities, consensual inductions, and unchallenged deductions.

Nevertheless, the sociologist, or the historian, will always want to enquire how that situation came into being and what gives it its stability. That inquiry is often difficult. As Harry Collins puts it, stable knowledge is like a ship in a bottle.[47] Looking at it as it is now, at least as naive observers, we cannot imagine how it was ever just a collection of sticks. How the ship got into the bottle can no longer be seen; yet human agency did put it in there. And that is perhaps the ultimate reason why we need the history of technology. Not to settle priority disputes, not to satisfy antiquarian curiosity, not for celebration, but because only through history can we find how the ship got into the bottle, how the technological artefacts and the technological knowledge we take for granted became takeable for granted.

---

[46] For examples, see MacKenzie, *Inventing Accuracy*.

[47] Harry Collins, "The seven sexes: a study in the sociology of a phenomenon, or the replication of experiments in physics", *Sociology* (1975), **9**: 205–24.

# Index

Milton Keynes UK
Ingram Content Group UK Ltd.
UKHW031145141024
449569UK00024B/1050